AF707334

Coordination Compounds for Pharmaceuticals and Drug Delivery Systems

(*Volume 1*)

Edited by

Abhay Nanda Srivastva
University Department of Chemistry
Babasaheb Bhimrao Ambedkar Bihar University
Muzaffarpur, India

Coordination Compounds for Pharmaceuticals and Drug Delivery Systems

(Volume 1)

Editor: Abhay Nanda Srivastva

ISBN (Online): 979-8-89881-429-8

ISBN (Print): 979-8-89881-430-4

ISBN (Paperback): 979-8-89881-431-1

© 2026, Bentham Books imprint.

Published by Bentham Science Publishers Pte. Ltd. Singapore, in collaboration with Eureka Conferences, USA. All Rights Reserved.

First published in 2026.

BENTHAM SCIENCE PUBLISHERS LTD.

End User License Agreement (for non-institutional, personal use)

This is an agreement between you and Bentham Science Publishers Ltd. Please read this License Agreement carefully before using the ebook/echapter/ejournal (**"Work"**). Your use of the Work constitutes your agreement to the terms and conditions set forth in this License Agreement. If you do not agree to these terms and conditions then you should not use the Work.

Bentham Science Publishers agrees to grant you a non-exclusive, non-transferable limited license to use the Work subject to and in accordance with the following terms and conditions. This License Agreement is for non-library, personal use only. For a library / institutional / multi user license in respect of the Work, please contact: permission@benthamscience.org.

Usage Rules:

1. All rights reserved: The Work is the subject of copyright and Bentham Science Publishers either owns the Work (and the copyright in it) or is licensed to distribute the Work. You shall not copy, reproduce, modify, remove, delete, augment, add to, publish, transmit, sell, resell, create derivative works from, or in any way exploit the Work or make the Work available for others to do any of the same, in any form or by any means, in whole or in part, in each case without the prior written permission of Bentham Science Publishers, unless stated otherwise in this License Agreement.
2. You may download a copy of the Work on one occasion to one personal computer (including tablet, laptop, desktop, or other such devices). You may make one back-up copy of the Work to avoid losing it.
3. The unauthorised use or distribution of copyrighted or other proprietary content is illegal and could subject you to liability for substantial money damages. You will be liable for any damage resulting from your misuse of the Work or any violation of this License Agreement, including any infringement by you of copyrights or proprietary rights.

Disclaimer:

Bentham Science Publishers does not guarantee that the information in the Work is error-free, or warrant that it will meet your requirements or that access to the Work will be uninterrupted or error-free. The Work is provided "as is" without warranty of any kind, either express or implied or statutory, including, without limitation, implied warranties of merchantability and fitness for a particular purpose. The entire risk as to the results and performance of the Work is assumed by you. No responsibility is assumed by Bentham Science Publishers, its staff, editors and/or authors for any injury and/or damage to persons or property as a matter of products liability, negligence or otherwise, or from any use or operation of any methods, products instruction, advertisements or ideas contained in the Work.

Limitation of Liability:

In no event will Bentham Science Publishers, its staff, editors and/or authors, be liable for any damages, including, without limitation, special, incidental and/or consequential damages and/or damages for lost data and/or profits arising out of (whether directly or indirectly) the use or inability to use the Work. The entire liability of Bentham Science Publishers shall be limited to the amount actually paid by you for the Work.

General:

1. Any dispute or claim arising out of or in connection with this License Agreement or the Work (including non-contractual disputes or claims) will be governed by and construed in accordance with the laws of Singapore. Each party agrees that the courts of the state of Singapore shall have exclusive jurisdiction to settle any dispute or claim arising out of or in connection with this License Agreement or the Work (including non-contractual disputes or claims).
2. Your rights under this License Agreement will automatically terminate without notice and without the

need for a court order if at any point you breach any terms of this License Agreement. In no event will any delay or failure by Bentham Science Publishers in enforcing your compliance with this License Agreement constitute a waiver of any of its rights.

3. You acknowledge that you have read this License Agreement, and agree to be bound by its terms and conditions. To the extent that any other terms and conditions presented on any website of Bentham Science Publishers conflict with, or are inconsistent with, the terms and conditions set out in this License Agreement, you acknowledge that the terms and conditions set out in this License Agreement shall prevail.

Bentham Science Publishers Pte. Ltd.
No. 9 Raffles Place
Office No. 26-01
Singapore 048619
Singapore
Email: subscriptions@benthamscience.net

CONTENTS

FOREWORD

The book entitled "Coordination Compounds for Pharmaceuticals and Drug Delivery Systems (Volume 1)" Edited by Dr. Abhay Nanda Srivastva provides a detailed insight about the significance of coordination compounds for the treatment of many diseases. The book is a compilation of valuable chapters describing the prophylactic properties of coordination compounds in drug delivery to cure many lethal diseases. Increasing threat of diseases is momentous issue across the globe. After the Covid-19 pandemic, it has been observed that there is need to aggrandize the preparation of more efficient drugs. Coordination compounds have bolstered the drug delivery system as they leveraged the therapeutic ability of drugs. A plethora of drugs have covered the pharmaceutical industry as many class of diseases can be cured by suitable mechanism/properties viz. antibacterial, antifungal, anti-alzheimers, antidiabetic and many more provided by coordination compounds. Lethal diseases like cancer can be cured by metallic drugs. These classes of compounds possess diversity in chemical and structural against lethal microorganism. The excellent drug delivery of the cis-platin has gained credence and many metal complex based drugs are under trial. Synergistic approach of pharmaceutical chemists across the globe to develop more efficient drug have augmented the applicability of coordination compound based drugs. The book is the recapitulation of intriguing drug delivery ability of the coordination compounds. Research scholars as well as undergraduate and postgraduate students of chemistry, biochemistry, and pharmacy along with chemical and pharmaceutical industrialist will be most benefitted from this volume of book. I wish the editor, authors and publication team all the success to this book.

Netra Pal Singh
Department of Chemistry
Dr. H.S. Gour University
Sagar, Madhya Pradesh
India

PREFACE

Chemical moieties occupied with medicinal values are the most important molecules as they cure and control many diseases to make the survival of living beings easier. Medicinal chemists always try their best to produce high quality of medicinal molecules from different branches of chemistry. Coordination compounds, among the various kinds of chemicals, have their unique importance as drug like molecules. Bioactive coordination compounds possess unique mode of drug actions due to versatility of ligands and metal ions attached together in different mode of coordination. Among the commercially available frontline drugs for curing many lethal diseases, metal based pharmaceutics like Cisplatin, Carboplatin, Oxaliplatin, Salvarsan, Auranofin, Bactiracin zinc salt *etc.* also cover an enough part of global pharmaceutical industries. Diagnosis capital is also roofed with Gadobutrol and Gadodiamide, Gd(III) complexes, as the commercial MRI contrasting agent. After taking an insight of these enthralled medicinal, diagnostic and theranostic appeal of metal complexes, pharmaco-researchers are focusing on the developments of potent metallic therapeutic moiety. The first volume of the book entitled "Coordination compounds for Pharmaceuticals and Drug delivery systems" is a compilation of advancements and research findings in the area of novel metal complex designing and bioactive applications from antibacterial to anticancer. An exhaustive discussion on bioactivity of metal complexes on life style generated diseases like diabetes is also focused in a chapter of the book. The advantages of coordination compounds in different drug delivery systems have also been discussed in the chapters embedded here.

I hope this compilation will be advantageous for all the personnel belonging to drug chemistry, pharmaceutical chemistry, coordination chemistry and bio-inorganic chemistry from academia to research to industries in a very simple and comprehensive style of service as a book.

Abhay Nanda Srivastva
University Department of Chemistry
Babasaheb Bhimrao Ambedkar Bihar University
Muzaffarpur, India

DEDICATION

Dedicated to my respected mother *Mrs. Sandhya Srivastava* and loving kids of the family; *Shivansh, Rudransh, Divyansh, Chitransh & Ashutosh*

List of Contributors

A.P. Mishra	Bio-inorganic Research Lab, Department of Chemistry, Dr. Hari Singh Gour Vishwavidyalaya (A Central University), Sagar, India
Anamika	Department of Chemistry, Harcourt Butler Technical University, Kanpur, India
Anindra Sharma	Department of Chemistry, Lalit Narayan Mithila University, Darbhanga, Bihar, India
Anjali Upadhyay	Department of Chemistry, D.D.U. Gorakhpur University, Gorakhpur, Uttar Pradesh, India
Aparna Tripathi	Department of Chemistry, Siddharth University, Siddharth Nagar, Uttar Pradesh, India
Ashish R. Dwivedi	GITAM School of Pharmacy, GITAM University, Hyderabad, India
Chote Lal Yadav	Department of Chemistry and Environmental Science, Madan Mohan Malaviya University of Technology, Gorakhpur, Uttar Pradesh, India
Deepak Tomar	Department of Chemistry, University of Delhi, Delhi, India Department of Chemistry, RK(PG) College, Shamli, Uttar Pradesh, India
Divyakshi Arya	Department of Chemistry, D.D.U. Gorakhpur University, Gorakhpur, Uttar Pradesh, India
Gulshan A. Khan	Department of Chemistry, D.D.U. Gorakhpur University, Gorakhpur, Uttar Pradesh, India
Khushi	Department of Chemistry, Siddharth University, Siddharth Nagar, Uttar Pradesh, India
Krishna K. Manar	Department of Chemistry, Faculty of Science, University of Allahabad, Prayagraj, Uttar Pradesh, India
Madhuri Chaurasia	Department of Chemistry, University of Delhi, Delhi, India
Mahima Pandey	Department of Chemistry, Siddharth University, Siddharth Nagar, Uttar Pradesh, India
Manisha	Department of Chemistry, SRM Institute of Science and Technology, Delhi-NCR Campus, Modinagar, India
M.L. Sharma	Central Department of Chemistry, Tribhuvan University, Kirtipur, Kathmandu, Nepal
Mohan Kumar	Department of Chemistry, M.M.H. College, Ghaziabad, Uttar Pradesh, India
Najiya Khatoon	Department of Chemistry, Siddharth University, Siddharth Nagar, Uttar Pradesh, India
Nisha Saxena	Department of Chemistry, M.R.M. College (Lalit Narayan Mithila University), Darbhanga, Bihar, India
Noopur Srivastava	Department of Chemistry & Biochemistry, Sharda School of Engineering & Sciences (SSES), Sharda University, Greater Noida, Uttar Pradesh, India
Pallavi Jain	Department of Chemistry, SRM Institute of Science and Technology, Delhi-NCR Campus, Modinagar, India

Priya Yadav	Department of Chemistry, Amity Institute of Applied Sciences, Amity University, Noida, Uttar Pradesh, India
Ranjana	Department of Chemistry, Siddharth University, Siddharth Nagar, Uttar Pradesh, India
Sarvesh K. Pandey	Department of Chemistry, D.D.U. Gorakhpur University, Gorakhpur, Uttar Pradesh, India
Shweta Singh	Department of Chemistry, D.D.U. Gorakhpur University, Gorakhpur, Uttar Pradesh, India
Shweta Srivastava	Department of Chemistry, Siddharth University, Siddharth Nagar, Uttar Pradesh, India
Som S. Dubey	Department of Chemistry, D.D.U. Gorakhpur University, Gorakhpur, Uttar Pradesh, India
Sulekh Chandra	Department of Chemistry, Zakir Husain Delhi College, University of Delhi, New Delhi, India
Sunil K. Srivastava	Department of Chemistry, Siddharth University, Siddharth Nagar, Uttar Pradesh, India
Vatsala Kumari	Department of Chemistry, Faculty of Science, University of Allahabad, Prayagraj, Uttar Pradesh, India
Vinayshri Tripathi	Bio-inorganic Research Lab, Department of Chemistry, Dr. Hari Singh Gour Vishwavidyalaya (A Central University), Sagar, India

CHAPTER 1

Advances in Coordination Compounds for the Drug Delivery System in the Human Circulatory System

Sunil K. Srivastava[1,*], **Mahima Pandey**[1], **Shweta Srivastava**[1], **Aparna Tripathi**[1], **Najiya Khatoon**[1], **Khushi**[1] and **Ranjana**[1]

[1] *Department of Chemistry, Siddharth University, Siddharth Nagar, Uttar Pradesh, India*

Abstract: Background: Drug delivery is a significant and evolving research area that relies on accurately quantifying drugs, particularly when developing drug delivery systems and understanding the pharmacokinetics and pharmacodynamics of actives in coordination with metals. Metals, acting as catalysts or reactants in the determination process, play a vital role in drug quantification. This study underscores the crucial role of metals in drug quantification and their coordination with actives, providing valuable insights into drug delivery and recent advances in drug applications. By emphasizing the role of metals in drug quantification, the audience can feel more informed and knowledgeable about this crucial aspect of drug delivery.

Methodology: This study utilizes Artificial Intelligence (AI) to review 1199 articles from the search engine “Advances in Coordination Compounds for the Drug Delivery System in the Human Circulatory System.” The AI tools were instrumental in analyzing and synthesizing the vast data on the 'Drug Delivery System in the Human Circulatory System.' Specifically, AI was used to identify relevant articles, extract essential information, and analyze trends in the research.

Key Findings and Future Prospects: Transforming the old drug delivery system from its traditional form into a unique delivery system is a process and a journey of rejuvenation. It can significantly enhance compliance, safety, and efficacy, breathing new life into an old drug molecule. The various technological advances in unit operations, such as drying, filtration, and mixing, are crucial and the backbone of this transformation, modifying and improving drug formulations. The potential of advanced drug delivery systems utilizing lipidic, proteic, and polymeric technologies **is of immense importance**. These systems have the power to revolutionize drug delivery, offering a hopeful and exciting future for pharmaceuticals. They provide sustained drug delivery with better body distribution, protection from harsh external environments, and avoidance of drug clearance. **Hence,** a rapid drug delivery system **could be effective in case of a fast**-dissolving tablet, a sustained system could be a transdermal patch, and a highly efficient system could be a targeted drug delivery system. The

* **Corresponding author Sunil K. Srivastava:** Department of Chemistry, Siddharth University, Siddharth Nagar, Uttar Pradesh, India; E-mail: sunil44@suksn.edu.in

Abhay Nanda Srivastva (Ed.)
All rights reserved-© 2026 Bentham Science Publishers

potential of these new systems is inspiring and a call for further exploration and innovation in the field. By stressing the importance of further exploration and innovation, the **researchers** can feel optimistic about the future of pharmaceuticals.

Conclusion: Technology-based drug delivery systems play a crucial role in facilitating and storing drug molecules in suitable forms for administration. However, their scope extends beyond small molecules. The potential of cell therapies is a beacon of hope in the field, with their promise of a sustained source of complex biologics. These therapies have the potential to revolutionize drug delivery, breaking innate biological barriers and creating natural responses within the system.

Keywords: Coordination compounds, Circulatory system, Drug, Metal, Pharmaceutical.

INTRODUCTION

The recent advances in the innovation of coordination compound research have underscored the significance of drug delivery mechanisms, particularly their high selectivity, which inspires confidence in their effectiveness. The coordination compound's role in targeting various diseases, from common ailments to more complex conditions, is a testament to its medical potential. The unique and distinct structural order of drugs based on coordination compounds, a key feature, is crucial and fascinating. This unique structural order of coordination compounds is vital to the human body's biological systems. The hemoglobin (a metal complex of Fe^{2+} ions) in human blood is a key player in oxygen transport and is crucial to the functioning of the human body's metabolism [1]. It was observed that coordination compounds have been instrumental in developing novel derivatives with therapeutic potential, a field ripe with possibilities that should pique the interest and curiosity of researchers.

The medicine can be easily encapsulated in nanomaterials and used as transporters for treatment, allowing for prolonged and well-organized drug administration in the human body. In the recent past, the transformative power of nanoparticles has reshaped contemporary medicine. By enhancing bioavailability, which refers to the proportion of a drug or other substance that enters the circulation when introduced into the body and so is made available for use or excretion, improving targeting, and modulating drug release mechanisms, these innovations have opened up new possibilities. Traditional drug delivery methods, with their poor therapeutic efficacy due to rapid elimination, inadequate solubility, and systemic toxicity, are being re-imagined. Nanomaterials, with their high surface area-to-volume ratio, configurable surface properties, and controlled drug delivery applications, are leading this revolution [2].

This is one of the new innovative areas that must be explored for drug delivery at this research stage. The analysis of experimental data indicates that nanoparticles are designed for the specific targeted tissues by coupling with targeting ligands and are designed to contain numerous agents. The research suggests that the most popular drug delivery carriers are preferably based on dendrimers, liposomes, organic polymeric, and inorganic particles. These innovative methods are crucial for designing a suitable drug delivery system at the nanoscale for the coordination of polymer particles encapsulating desired components. The technique, which involves a coordination polymerization, is further followed by quick precipitation, and it offers several advantages over existing approaches. It shows its importance during application through entrapping organic dyes, quantum dots, and magnetic nanoparticles into blue fluorescent spheres created through the coordination of Zn^{2+} with 1,4-bis(imidazol-1-ylmethyl) benzene (bix) organic ligands [3]. This stretches the significance of coordination compounds in medical sciences.

This research has the potential to significantly impact drug delivery systems, offering an efficient and controlled method of drug administration. The study indicates that the unique characteristics of nanomaterials enhance the ability of medical science to entrap a wide variety of substances, which facilitates fast and efficient medical treatment. Researchers reported in their study that coordination polymer nanospheres function as better drug-encapsulating agents and can be utilized as novel functional carriers for targeted medicine delivery to specific cells. The potential impact of this research on the drug delivery systems field sparked their curiosity and eagerness to learn more about the hopeful and exciting future it presents. There are numerous sources of natural bioactive molecules, which are essential for a comprehensive understanding of their function. Several selected sources of natural bioactive molecules are available in nature, as illustrated in Fig. (**1**). These bioactive molecules possess numerous clinical applications, including anticancer, antioxidant, immunomodulatory, antidiabetic, anti-obesity, neuroprotective, and anti-inflammatory properties.

Ma and Moulton [4] have studied innovative advances in discrete coordination complexes and polymers, which facilitate upgrading their applications for drug delivery systems. They also emphasized and listed the significance of these compounds in modern research for modern drug delivery systems. Recent material science and crystal engineering research has allowed researchers to promote coordination complexes as functional materials. The most popular applications are catalysts, magnetic, non-linear optical, and porous materials [4].

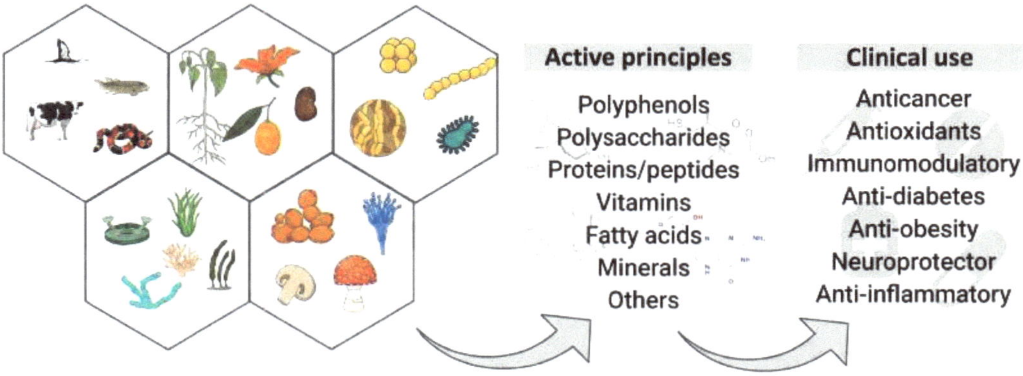

Fig. (1). Diversified source of a few selected natural bioactive molecules.

In their research, Khan *et al.* [5] highlighted the potential of nanotherapeutic strategies for treating Huntington's Disease (HD), a significant medical concern for society. The methods adopted by the above researchers for the specific target and delivery of therapeutic agents to selected sites display considerable potential in treating the above disease by addressing intricate molecular pathways [5]. The other research scholar reported a significant impact of this method for resolving the medical concern of the above disease.

Dai *et al.* [6] explored the advantages of porphyrin-based supramolecular nanomaterials for phototherapy. These phototherapies, including photodynamic therapy and photothermal therapy, have emerged as novel clinical techniques. Their low-toxic side effects, lack of drug resistance, and strong immune activation effects provide a sense of reassurance and confidence in their potential, ensuring a secure and confident approach to medical treatment.

Kothawade and Shende [7] reported in their research work about coordination-bonded stimuli-responsive delivery systems of chemical actives with metal in pharmaceutical applications. It is a highly novel and intriguing approach to new drug delivery in the human body. They reported in their study that stimuli-responsive drug delivery systems exhibit active responses to pH, enzyme, redox, magnetic, thermal, and light stimuli. It is further noticed through a reversible coordination bond with metal ions and ligands. Numerous ligands show massive potential in binding with metal ions and broad applications in drug delivery; a few popular ligands include ethylenediamine tetraacetic acid and azobenzene [7].

The potential of coordination compound polymer in drug delivery offers a promising future for pharmaceuticals, instilling a sense of optimism in the audience. The researchers reported in their study that drug delivery systems in pharmaceuticals depend on the nature of bonds, such as imine, metal-ligand coordination, hydrogen bonds, and pi-pi stacking, to store and release drugs in response to specific triggering agents [7]. Due to its high surface areas, pH-sensitive behaviors, complexes, and coordination polymer networks, it explores personalized drug release.

Researchers have delved into the possibilities of a new approach in coordination chemistry, a significant leap from the perspective of advances in medical science. This new approach, which includes understanding the mechanism of anti-tumor activity of selected heterometallic complexes [8 - 10], holds great promise and importance in the field of drug delivery systems.

In the case of therapy for a patient who has liver cancer, the experimental results from supramolecular assemblies suggest satisfactory results in precise drug release in response to stimuli from the tumor microenvironment. The concept of drug supply to specific tissue compartments within cells, known as cellular and intracellular targeting, is gaining crucial traction [11]. The research community worldwide indicates that nanocarriers possess a promising technology that can effectively deliver drugs to sites where drug penetration is challenging due to anatomical barriers. The blood-brain barrier, for instance, poses a significant and complex challenge to drug delivery, as it selectively restricts the entry of most drugs into the brain. However, the potential of nanocarriers for the effective delivery of medicines to these challenging sites instills confidence in the future of drug delivery systems and a sense of triumph over obstacles.

Types of Drug Delivery Systems

There are two identified and most accepted current-stage drug delivery systems; the first is active, and the second is passive. The specific marker expressed in the diseased cells is under the active mode of delivery [12]. This targeting mechanism can be highly proficient by allowing a molecule to bind to overexpressed folate receptors in diseased cells [11]. CA-125 is one of the biomarkers overexpressed in ovarian cancer and can be targeted for active targeting mechanisms. It is observed that active targeting includes antibodies that bind to specific proteins on cancer cells or ligands that bind to receptors on the surface of the cancer cells.

In the passive mode mechanism, the growth of the biocompatible polymer at the site of infected cells depends mainly on its size. In this case, the polymers may be extravasated due to leaky junctions in the vessels, further allowing them to reach the diseased site [13]. The identified drug delivery route depends on various factors, a few of which are elaborated in another section of this article.

Dosage

A dose of medicine consists of Active Pharmaceutical Ingredients (API) and excipients or additives. Excipients play a crucial role in medicine formulation and serve as coloring agents, suspending agents, flavoring agents, solubilizing agents, and antioxidants [14]. The selection of excipients is based on the purpose of the medicine. For instance, lactose is often included to increase the size of the tablet, as an accurate dose of medicine is too small to handle without filler.

Understanding the classification of drugs based on biopharmaceutics is not just a crucial aspect of pharmaceutical knowledge but a key to mastering drug formulation and administration, empowering you to make informed decisions. Drugs are categorized into four major classes, each with its unique characteristics and implications for drug use.

- Class I: Class I drugs show high permeability and solubility compared to other classes of drugs. Their absorption rate is usually more crucial/important than excretion, *e.g.*, metoprolol, paracetamol, *etc*.
- Class II: Class II drugs show high permeability but low solubility compared to other classes of drugs. Further, such drugs, such as glibenclamide and aceclofenac, have restricted or limited bioavailability due to their rate of solvation.
- Class III: Class III drugs generally possess low permeability and high solubility compared to other classes. This may be due to their low permeability; thus, these drugs exhibit limited absorption, for example, Cimetidine.
- Class IV: Class IV drugs show poor absorption in the intestine, which may be due to low permeability and solubility. Hence, such a class of drugs has poor bioavailability with high variability in comparison to other classes of drugs, *e.g.*, Bifonazole.

Nanocarriers

Nanocarriers are ideal for drug delivery systems, which are rich with innovation and possess a significant role in hitting the target. Each nanocarrier offers a unique advantage and challenge, making the field of drug delivery a fascinating area of research/study, from liposomes to carbon nanotubes [15]. Nanocarriers are one of the most versatile research areas, resulting in fruitful innovations in drug delivery systems. The few selected nanocarriers are listed below, considering their importance and significance.

- Liposomes
- Dendrimers

- Exosomes
- Nanoparticles
- Nanosphere or nanocapsule
- Solid-lipid nanoparticles
- Nanofibers
- Polymersomes
- Self-assembled polymeric micelles
- Carbon nanotubes
- Nano-emulsion
- Hydrogels, *etc.*

The Routes of the Drug Administration

Identifying a suitable drug administration route is a crucial part of pharmaceutical practice. It requires careful consideration and attention to detail, as determined by the target site, treatment duration, and the drug's physicochemical attributes. An in-depth understanding of these factors is crucial for the effective implementation of drugs [16]. The following methods are a few selected popular routes of drug administration, which possess significant importance in the modern age of medical science and a bright future in the area of medical science.

- Buccal: These methods hold the drugs inside the cheek.
- Sub-lingual: In this method, drugs are placed below the tongue.
- Enteral: In such cases of drug administration, the drug is delivered directly into the stomach or intestine.
- Inhalable: In this type of drug administration, drugs are breathed in through a tube or mask.
- Nasal: In this method of drug administration, drugs are given into the nose by spray or pump.
- Ophthalmic: The drugs are directly given into the eye by drops, gel, or ointment.
- Otic: In this case, drugs are administered by drops into the ear.
- Rectal: In such cases, drugs are inserted into the rectum.
- Vaginal: In such types of drug administration, drugs are inserted into the vagina.
- Topical: In this case, drugs are applied to the skin.
- Transdermal: In this method, drugs are given through a patch on the skin.
- Infused: In such cases, drugs are injected into the vein with an IV line and slowly dripped in over time.
- Intramuscular: In such cases, drugs are injected into the muscle with a syringe.
- Intravenous: In this case, drugs are injected into a vein or through an IV line.
- Subcutaneous: In this method, drugs are injected just under the skin.

This critical review is about advances in coordination compounds for the drug delivery system in the human circulatory system. An advanced coordination compound offers a broad scope of applications for coordination compounds, facilitating safe and secure drug delivery in human biological systems.

METHODOLOGY

This critical review was initiated by searching this unique topic in the "Science Direct" and other search engines. At the initial stage, searching for "Coordination Compound" in the search engine, focusing on its application in drug delivery systems, returned 408,002 articles. The search for "Coordination Compound for Drug Delivery System" yielded 29,843 articles at the next stage. A further search of "Advances in Coordination Compounds for the Drug Delivery System in the Human Circulatory System" in the "Science Direct" search engine, focusing on recent developments, gave 1199 articles. From these articles, we found that the use of coordination compounds in drug delivery systems has been a growing area of research, with a significant increase in publications in recent years. These articles are classified year-wise as:

2025 [10 articles],
2024 [162 articles],
2023 [92 articles],
2022 [86 articles],
2021 [84 articles],
2020 [75 articles],
2019 [53 articles],
2018 [38 articles],
2017 [52 articles],
2016 [32 articles],
2015 [21 articles],
2014 [34 articles],
2013 [31 articles],
2012 [24 articles],
2011 [20 articles],
2010 [23 articles],
2009 [21 articles],
2008 [25 articles],
2007 [23 articles],
2006 [18 articles],
2005 [21 articles],
2004 [18 articles],
2003 [7 articles],

2002 [9 articles] and
2001 [10 articles].

The more recent research publications have indicated the significance of coordination compounds in drug delivery systems. Artificial Intelligence (AI) facilitates a quick and efficient review process, which enhances the possibility of a highly efficient critical review. This also ensured a comprehensive review of the current state of research in this field, reassuring the audience about the thoroughness of the research process and building trust in the study's findings, thereby enhancing their confidence in the validity and reliability of the research.

RESULTS AND DISCUSSION

Recent research has revolutionized the pharmaceutical industry by advancing innovative coordination compounds for drug delivery mechanisms. Further, integrating Artificial Intelligence (AI) and Machine Learning (ML) techniques has opened up numerous opportunities for the scientific community, particularly in the precise treatment of complex cancer cases. These advanced coordination compounds enable drug delivery directly to the target, sparing healthy neighbouring cells of the tumour. The role of AI and ML in optimizing drug delivery systems, from predicting drug behaviour in the body to designing more effective drug delivery mechanisms, must be addressed [17 - 22]. The researchers utilized this modern, innovative technique to resolve challenges raised during drug administration. The results indicate satisfactory performance for these new innovative technologies of drug administration.

Innovation and Phase Development in Drug Delivery Systems

The drug delivery system is a mechanism in which drugs are delivered to the target diseased cell for effective and efficient treatments without harming neighbouring cells. The history of the drug delivery system, classified into four primary categories, provides a comprehensive understanding of the year-wise development of the drug delivery mechanism [23]. This knowledge equips us with a deeper understanding of drug delivery systems. The history of the development of drug delivery systems is classified into four significant phases of development, and details are given below:

- First Phase of Development—Basics of Controlled Release (1950-1980): In this phase of drug delivery system development, sustained-release capsule technology and oral and transdermal controlled-release technology were popular. Polymeric biomaterials are preferably used in wax coating and transdermal patches, utilizing dextran and silicon rubber.
- Second Phase of Development- Innovative delivery system (1980-2010): In this

phase, bright polymer and hydrogel were developed, a stimulus-response system was implemented, and delivery of peptides and proteins (PEG) was implemented with an injectable or implantable depot system. An intelligent delivery system can respond to specific stimuli, such as changes in pH or temperature, to release the drug at the desired site without harming healthy cells [23].

- Third Phase of Development—Modulated Drug Delivery System (2010-2024): In this phase, nanomedicine development propagated, including diseases other than cancer, targeted delivery, and formulation, aiming to overcome physicochemical and biological barriers. Quantum dots, gold nanoshells, mesoporous silica, metal oxide, and carbon nanotubes are used as drug delivery instruments for medical treatment. This phase still needs wider-scale applications worldwide and more in-depth research.
- Fourth Phase of development—Future Prospective (2024 to date): This is the youngest and most innovative phase, with the facility of AI and ML techniques. In this phase, research will be carried out for nanomedicine refinement. Medicine's precision and accuracy will be enhanced to reduce the leading cause of death. The formulation will upgrade the body's defense system and revolutionize medical science worldwide. This phase of the drug delivery system may completely change medical science due to its exact and accurate results.

It is observed that further research needs to be conducted on the evolution of macromolecule delivery systems. These prospects in nanomedicine are promising and revolutionary, offering a glimpse into the potential advancements in drug delivery systems. This article emphasizes the target-oriented delivery systems that target specific structures of desired cells. This new concept was adopted by the scientific community recently and needs more research for better results. This concept is stimulating and inspiring when considering the physicochemical features of the frequently marketed biotechnological macromolecules (peptides, proteins, antibodies, and nucleic acids) [24, 25]. Nanocarriers, such as nanoparticles, can transport ocular drugs to exact target sites and promise to revolutionize therapy through their high accuracy and precision [24].

Nucleic acid ligands, also known as aptamers, are a popular class of macromolecules used in several novel nano-biomedical applications [25]. Huxford *et al.* [26] studied and elaborated on metal-organic frameworks, a new hybrid material class composed of suitable metal ions and organic bridging ligands. These have emerged as new, upcoming platforms for drug delivery, offering potential hope for the future of drug delivery systems.

Horcajada *et al.* [27] first studied and described the adsorption and release of ibuprofen. Meanwhile, Rieter *et al.* [28] have structured a Pt-based drug at the nanoscale, using it as one building block for creating the framework. The degradable nanoparticle formulation facilitates the design of NCPs as effective delivery vehicles for various biologically and medically essential cargoes, such as therapeutic and imaging agents [28]. The recent surge in the application of AI and ML techniques in medical science plays a crucial and reassuring role in minimizing challenges and improving efficiency, particularly in drug delivery systems [29, 30]. These technologies can be used for drug design, predicting drug interactions, optimizing drug formulations, and personalizing drug dosages, thereby revolutionizing the field of drug delivery. This technological advancement is a beacon of hope, promising a brighter future for medical science and drug delivery.

Concern Zone of the Drug Delivery Mechanism

According to the FDA, an *active drug* is designed for the diagnosis, treatment, mitigation, or prevention of disease. It is a method of administering medication to a patient that increases the drug concentration in the affected body area and is a crucial technique [31 - 33]. The drug delivery system plays a vital role in our field, as the direct clinical use of active drug substances (APIs) is rare due to various reasons, such as the difficulty of handling and accurately dosing very potent drugs (*e.g.*, low mg and μg doses) [32, 34]. The challenges in direct drug administration, such as the impracticality and infeasibility of administering drugs into body cavities (rectal, vaginal) due to degradation and local irritations, highlight the essential role of innovative drug delivery systems. These systems are not just a convenience but a necessity in our work. The need for innovative drug delivery systems is further underscored by the challenges in direct drug administration, such as the impracticality and infeasibility of administering drugs into body cavities (rectal, vaginal) due to degradation and local irritations. These challenges can lead to inconsistent drug delivery and reduced patient compliance, emphasizing the need for more effective delivery systems.

Pharmacokinetics of the Drug Delivery System

Pharmacokinetics is the study of the rate of transport of drugs into the human body's circulatory system. It includes all metabolic activities within and outside the body, including the time spent by the drug during absorption, distribution, metabolism, and excretion [35, 36]. These processes are crucial as they determine the concentration of a drug at its site of action, duration of action, and intensity of action. A drug has four major pharmacokinetic phases during its passage through its over-designed routes. These are as given below:

- Absorption: The absorption process facilitates the drug's movement from its administration site to the bloodstream. The rate and extent of drug absorption depend on several factors, such as the route of administration (*e.g.*, oral, intravenous, intramuscular), the drug's physicochemical properties (*e.g.*, solubility, stability), the type of formulation (*e.g.*, tablet, capsule, solution), and, notably, drug-food interactions [37]. These interactions will significantly affect the absorption of a drug, underscoring the importance of considering them in pharmacokinetics [37, 38]. Thus, this is one of the crucial steps of pharmacokinetics, which significantly influences the efficiency of medicine in our circulatory system.

- Distribution: The reversible transfer of a drug between the blood and the body's extravascular fluids and tissues is a pivotal phase in pharmacokinetics. Drug distribution, which determines the amount of drug reaching target sites compared to the rest of the body, is a crucial factor in drug efficacy as well as toxicity [39]. Factors such as blood flow, lipophilicity, molecular size of the drug, and binding affinity of the drug with plasma proteins significantly influence drug distribution, underscoring its importance [39, 40].
- Metabolism: The drugs are metabolized into inactive or less active components before being absorbed into the systemic circulation. The optimum dosing is essential for oral administration, as it is significantly reduced before reaching the bloodstream [41]. The dosing varies according to the mode of administration of the drug in the human body. It can't be the same for all types of administrative modes.
- Excretion: Drug excretion is a complex process that removes unchanged drugs or metabolites from the body. It involves many different routes of excretion, including urine, bile, sweat, saliva, tears, milk, and stool [42]. The complexity of drug excretion underscores the need for a comprehensive understanding of pharmacokinetics in drug development and clinical practice [42]. The properties of medicine and the mode of drug administration influence the quantity of drugs released during excretion. This is also important considering the toxic influence of the drugs in the environment during and after excretion.

The half-life of the Drug ($t_{1/2}$)

The elimination half-life or biological half-life ($t_{1/2}$) of a drug is the time it takes for the amount of the drug to become half (50%) of the initial concentration. This concept is crucial and essential in pharmacokinetics as it facilitates the assessment of the dosing frequency and the time it takes for a drug to be eliminated from the body [43 - 46]. As per scientific communities, the half-life of a drug in a biological system can be defined as per the following equations: (1) and (2). All scientific communities accept these equations.

$$t_{1/2} = (0.7 \times V_d)/Cl \quad \textbf{(1)}$$

Where V_d is the volume of distribution, and Cl is clearance.

$$t_{1/2} = 0.693/K_t \quad \textbf{(2)}$$

Where K_t is the elimination rate constant.

Those drugs, which have short biological half-lives, require rapid dosing for a sustained and effective therapeutic response. The primary objective is to maintain the therapeutic blood level for the necessary periods. It needs to optimize the drug dose to enter systemic circulation at a rate comparable to its elimination rate. The experimental findings underscore the critical role of drug half-life in determining treatment efficiency and effectiveness, which varies due to several factors, including age, weight, concurrent medications, underlying medical conditions, and kidney and liver function, among others [47]. This variability presents a significant puzzle for pharmacologists and researchers, highlighting the urgency and significance of their research in understanding and addressing these challenges. The half-life of a drug can significantly impact the frequency and dosage of administration, the duration of treatment, and the potential for drug accumulation or toxicity, underscoring its importance in drug delivery systems.

In practice, understanding the pharmacokinetics of a drug, including its half-life, is essential to understanding the manner of effective treatment. The half-life of a drug directly influences the dosing, kinetics, and dissolution rate. It also facilitates our understanding of how the drug dissolves in the defined metabolism of species. This knowledge is essential when considering the variability in absorption, dissolution, and excretion rates across different species and age groups. It underscores the practical significance of understanding the pharmacokinetics of a drug, ensuring the optimum dose for treating infected organs/cells of human beings. Understanding the half-life of medicine and the metabolism of the human circulatory system, along with Understanding the half-life of medicine and the metabolism of the human circulatory system, along with prodder drug delivery, will revolutionize medical science and necessitate further research.

Coordination Compounds for the Drug Delivery System

Recent innovations in coordination compounds have enhanced drug delivery efficiency, revolutionizing medical science in various ways. Researchers working on the role of coordination compounds in drug delivery delineated the massive possibilities of their practical application. The few methods adopted by scientific communities still need more in-depth study for more efficient and satisfactory results. Efficient drug delivery needs to upgrade the efficiency of medicine and minimize the challenges raised over the healthy cells of organs located near infected cells.

The following table shows the application of a few selected coordination compounds for drug delivery systems in the human circulatory system (Table **1**). The following table summarises appreciable research work carried out by researchers worldwide in coordination with the compound application as a drug delivery carrier in medical science [1, 7, 16, 48 - 52]. The authors emphasized specific case studies, including the role of metal ion coordination compounds as anti-cancer agents [1, 7], silica-based drug delivery systems [49], emerging platelet-based drug delivery systems [50], nanoparticle-based drug delivery systems [51], and hydrogel-based drug delivery systems [52].

Table 1. The brief on the application of coordination compounds and other popular drug delivery methods. fulfills the therapeutic purpose.

Sr No.	Coordination Compound	Application in Drug Delivery System	References
1	Metal ions (Pt, Pd, Ru, Ni, Cr, Ag, and Au) coordination compounds	Recent data has played a crucial role in our understanding of the medical applications of metal coordination compounds. It has highlighted the significant role of Pt, Pd, Ru, Ni, Cr, Ag, and Au-based coordination compounds in addressing various medical issues, particularly cancer. The use of these metal ions, such as platinum (Pt), palladium (Pd), ruthenium (Ru), nickel (Ni), chromium (Cr), silver (Ag), and gold (Au), as anti-cancer drugs is a key finding. Furthermore, the metal complexes of Cr, Au, and Cu ions have shown excellent anti-bacterial activity, further enriching our understanding of these compounds.	[1]
2	Disuccinatocisplatin(c,c,t-$Pt(NH_3)_2Cl_2(OH)_2$) synthesis	*Cisplatin*, a platinum-based chemotherapy agent, is widely used to treat various sarcomas, carcinomas, lymphomas, and germ cell tumors. Disuccinatocisplatin, a compound synthesized in the laboratory, undergoes a specific synthesis process. This process involves heating at 70 °C with vigorous magnetic stirring for five hours in the dark, followed by overnight stirring. After concentrating the mixture to ~10 mL, the product can precipitate at 4 °C over several hours. The product is then collected *via* vacuum filtration, washed with ice-cold water, ethanol, and diethyl ether, and vacuum-dried. The resulting product, a bright yellow powder, exhibits unique and impressive properties. These properties make Disuccinatocisplatin potentially useful. Yield: 1.911 g (85.8%)	[48]

(Table 1) cont.....

Sr No.	Coordination Compound	Application in Drug Delivery System	References
3	c,c,t-$Pt(NH_3)_2Cl_2(OOCCH_2CH_2CO_2H)_2$(DSCP)	These nanoscale coordination compound polymers are used for platinum-based anticancer drug delivery. In DSCP synthesis, the mixture of c,c,t-$Pt(NH_3)_2Cl_2(OH)_2$ (1.00 g, 3.00 mmol) and succinic anhydride (1.20 g, 12.0 mmol) in dimethyl sulfoxide (DMSO) (3.00 mL) was heated at 70 °C in the dark for 24hours with magnetic stirring. Lyophilization is used to remove the solvent. The pale-yellow product was recrystallized from acetone at approximately -20 °C, isolated *via* vacuum filtration, and washed with ice-cold acetone.	[48]
4	Silica-based drug delivery systems	The mesoporous silica nanoparticles have emerged as a pre-eminent platform in nanomedicine. revolutionizing it.	[49]
5	The emerging platelet-based drug delivery systems	Fig. (**2**): Platelet-based drug delivery System	[50]
6	The coordination bonds with metal ions (*e.g.* Zn^{2+}, Cu^{2+}, Fe^{2+}/Fe^{3+}) and ligands such as ethylenediamine, tetraacetic acid, and azobenzene	Stimuli-sensitive drug delivery systems exhibit active responses to pH, enzyme, redox, magnetic, thermal, and light stimuli through reversible coordination bonds.	[7]
7	Aptamers	Due to their high affinity and specificity for their defined target, aptamers undergo a versatile selection process and are remarkably easy to synthesize. Their small physical size further enhances their appeal, making them attractive molecules for targeting diseases or as therapeutics. Hence, aptamers have a wide range of applications in medical science.	[16]
8	Drug delivery systems based on nano-material	Drug delivery systems based on nano-material systems are entirely new and innovative areas of drug delivery in medical science. As studies suggest, fungal keratitis is a universally accepted blinding eye disease caused by fungi. It presents a significant societal challenge, specifically in Asian and African countries. India is one of the Asian countries suffering due to the challenge raised by fungal keratitis. A recent study shows that approximately ~10,50,000 cases were diagnosed annually, a common form of infectious keratitis, accounting for approximately 40–50% of microbial keratitis cases. This underscores the urgent need for further research in this area, considering the vulnerability of the vast population of Asia and Africa. The applications of coordination compounds as nanoparticles may resolve a certain level of this medical concern with high efficiency. The experimental results indicate satisfactory results, but more in-depth research and design are needed for the application.	[51]
9	The hydrogel-based drug delivery systems for diabetic bone defects	Fig. (**3**): Hydrogel-based drug delivery system	[52]

The above studies reported appreciable results that require further in-depth analysis and exploration of opportunities in drug delivery systems using coordination compounds (Figs. **2** and **3**). Several researchers have reported remarkable achievements in their studies. The in-depth study of coordination compound-based drug delivery systems yielded significant results. The drug delivery process is crucial in targeting various diseases in the human body due to its distinct functions and structural order. It plays a vital role in the biological circulatory systems, underscoring the importance of this research work in understanding human health [46]. The innovation in the area mentioned above might enhance the sustainable approach of modern medical science by reducing a certain level of medical waste, which is also considered a challenge for human civilization [53 - 58].

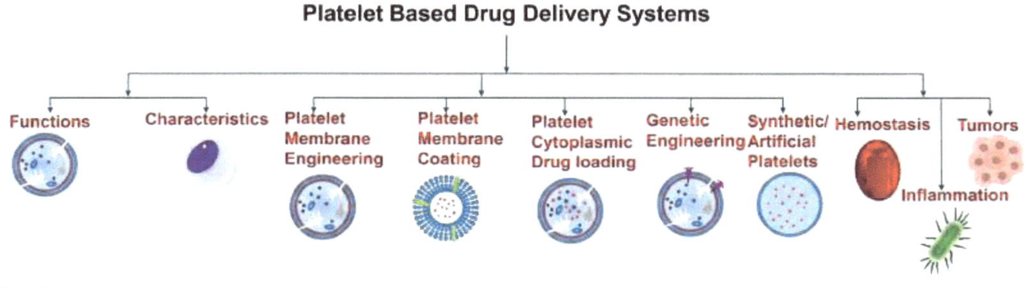

Fig. (2). Platelet-based drug delivery system.

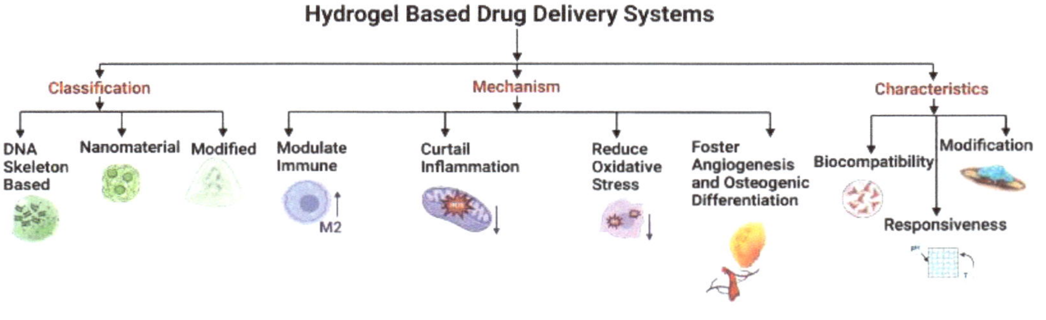

Fig. (3). Hydrogel-based drug delivery system.

The few popular coordination compounds with significant roles are listed: hemoglobin, a carrier of O_2 transportation in the blood, a metal complex of Fe(II). In addition, ongoing research in this field is uncovering new potential for these compounds. They have a rich historical significance, having been utilized for centuries to treat various diseases associated with humans due to their excellent antibacterial, anti-cancer, and antiviral properties. This historical context adds depth to the research and invokes a sense of respect and admiration for the long-standing contribution of coordination compounds to human health. The extensive investigation into coordination compounds of a few selected metal ions (Pt, Pd, Ru, Ni, Cr, Ag, and Au) reveals a wealth of potential in the fight against cancer. These compounds show promise as practical materials for anti-cancer drugs, inspiring further exploration and development in this critical area of research. Along with the compounds mentioned above, other more popular compounds need to be explored to serve human civilization, which will resolve the medical issue challenge of human beings. Further scope is always available when applying coordination compounds in medical science. This article provides a small exposure of the application and importance of the coordination compounds for the drug delivery system in the human circulatory system.

CONCLUSION

Drug delivery is a complex process that involves precisely targeting therapeutic molecules within cells without harming other healthy cells. It encompasses a series of intricate steps, including administration, distribution in biological organs, absorption by cells, and elimination (ADME). This complexity underscores the need for continued research and innovation in drug delivery methods. However, in general, old approaches, which depend on the systemic circulation of drugs, possess numerous challenges and side effects. It was observed that in older methods, more than 90% of the drug quantity in the systemic circulation was eliminated through renal clearance, leading to significant wastage of medicine that other sections of society could use and reducing the price of medicine.

An efficient drug delivery system helps to resolve typical medical cases efficiently. This underscores the urgent need for more effective drug delivery systems, particularly in complex medical cases like cancer. Coordination compounds, for instance, are proving to be practical and efficient tools for drug delivery in these cases without harming the human circulatory system or healthy organ cells. The controlled drug delivery process, which enhances drug solubility and stability, offers the selective delivery of drugs at a predictable rate and mechanism to specific organs, tissues, or cells. This precision in drug delivery not only enhances safety but also reassures the audience about the efficacy of the process.

Under pharmacokinetics, the study of drug movement into, through (in), and out of the body is a crucial aspect of drug delivery. These processes determine the concentration, duration of dissolution, chemical reaction, and intensity of a drug's action. Their significance lies in their ability to regulate the release of drugs, ensuring optimal therapeutic effects. The potential of nanocarriers is a promising area of research that highlights the exciting developments in drug delivery systems. It inspires the audience about the future of drug delivery systems. It reinforces their optimism about the possibility of these systems, instilling a sense of hope and anticipation for the future of pharmacology and drug delivery.

Research data indicate that coordination compounds have proven effective in resolving numerous medical issues. There is a clear need for more efficient, economically viable compounds for drug delivery systems. However, there is always room for improvement, and ongoing research in this area is crucial for advancing drug delivery systems. This emphasis on the need for more efficient coordination compounds not only instills a sense of urgency and importance in the audience but also highlights the continuous exploration and innovation in drug delivery systems that keep the field dynamic and promising, fostering a sense of commitment and dedication to the field of pharmacology and drug delivery. Coordination compounds are a present and future focus for efficient and economical methods in drug delivery systems, which require more in-depth study and exploration. This study concludes that the coordination compound may revolutionize medical science through innovative, selective, more efficient, and precise drug delivery methods. Significant research has been carried out; more in-depth information is needed for specific cases to resolve the issue efficiently. The approach and effort of the scientific community worldwide are appreciable, given its recent advances in efficient, innovative methods for drug delivery. It reduces significant medical challenges, helps lower indirect medical waste, and offers approaches to a sustainable environment.

DECLARATION

The authors declare that this research article has no financial or ethical conflicts of interest. They also declare no conflicts in their relationships with people in this university or any organizations that could inappropriately influence (bias) their research work. This research work is purely a critical review of research carried out worldwide, keeping in mind the welfare of society. This is an approach to identify new drug delivery methods for medical science.

FUNDING

No external sponsoring agency funded this research work. Siddharth University provided an ideal environment for this review research work.

ACKNOWLEDGMENTS

The authors thank the university's authority for providing valuable facilities for this review research work.

DATA AVAILABILITY

This is a thorough review of research work; no data is required, and appropriate citations for research work are provided at a suitable location in the article.

REFERENCES

[1] Kumar Singh V, Kumar Singh V, Mishra A, *et al.* Recent advancements in coordination compounds and their potential clinical application in the management of diseases: An up-to-date review. Polyhedron 2023; 241: 116485. [http://dx.doi.org/10.1016/j.poly.2023.116485]

[2] Islam S, Ahmed MMS, Islam MA, Hossain N, Chowdhury MA. Advances in nanoparticles in targeted drug delivery– A review. Results in Surface and interfaces 2025; 19:100529. [http://dx.doi.org/10.1016/j.rsurfi.2025.100529]

[3] Imaz I, Rubio-Martínez M, García-Fernández L, *et al.* Coordination polymer particles as potential drug delivery systems. Chem Commun (Camb) 2010; 46(26): 4737-9. [http://dx.doi.org/10.1039/c003084h] [PMID: 20485835]

[4] Ma Z, Moulton B. Recent advances of discrete coordination complexes and coordination polymers in drug delivery. Coord Chem Rev 2011; 255(15-16): 1623-41. [http://dx.doi.org/10.1016/j.ccr.2011.01.031]

[5] Khan S, Bano N, Ahamad S, Dar NJ, Nazir A, Bhat SA. Advances in nanotherapeutic strategies for Huntington's disease: Design, delivery, and neuroprotective mechanisms. Coord Chem Rev 2025; 522: 216206. [http://dx.doi.org/10.1016/j.ccr.2024.216206]

[6] Dai Y, Sun J, Fang L, Zhao J, Yu F, Wang B. Recent advancements of porphyrin-based supramolecular nanomaterials for phototherapy. Coord Chem Rev 2026; 547: 217121. [http://dx.doi.org/10.1016/j.ccr.2025.217121]

[7] S. Kothawade, P. Shende. Coordination-bonded stimuli-responsive drug delivery system of chemical actives with metal in pharmaceutical applications 510 (2024) 215851. [http://dx.doi.org/10.1016/j.ccr.2024.215851]

[8] Šaćirović E, Kolenović Serezlić M, Soldatović TV. A new approach in coordination chemistry from the perspective of the mechanism of antitumor activity of selected heterometallic complexes. J Coord Chem 2025; 1-18. [http://dx.doi.org/10.1080/00958972.2025.2558101]

[9] Krishnaveni A, Kalanithi M, Lazer A. Co(II), Ni(II), and Cu(II) complexes of Schiff base of anthracene: sensor and biological applications. J Coord Chem 2025; 78(18): 2067-80. [http://dx.doi.org/10.1080/00958972.2025.2539940]

[10] Zheng YJ, Xu ZX. Ligand-directed assembly of tetranuclear cobalt clusters: structural modulation *via* a lactate derivative and azacyclic linkers. J Coord Chem 2025; 78(18): 2023-36. [http://dx.doi.org/10.1080/00958972.2025.2538100]

[11] Yadav KS, Mishra DK, Deshpande A, Pethe AM. Levels of drug targeting. In: Tekade RK, Ed. Basic fundamentals of drug delivery. Cambridge, MA, USA: Academic Press 2019; pp. 269-305. Available from: https://www.sciencedirect.com/book/9780128179093/basic-fundamentals-of-drug-delivery [http://dx.doi.org/10.1016/B978-0-12-817909-3.00007-8]

[12] Attia MF, Anton N, Wallyn J, Omran Z, Vandamme TF. An overview of active and passive targeting strategies to improve the nanocarriers efficiency to tumour sites. J Pharm Pharmacol 2019; 71(8): 1185-98. [http://dx.doi.org/10.1111/jphp.13098] [PMID: 31049986]

[13] Shah A, Aftab S, Nisar J, Ashiq MN, Iftikhar FJ. Nanocarriers for targeted drug delivery. J Drug Deliv Sci Technol 2021; 62: 102426.
[http://dx.doi.org/10.1016/j.jddst.2021.102426]

[14] Chaudhari SP, Patil PS. Pharmaceutical excipients: A review. IJAPBC 2012; 1:21–34. Available from: https://scispace.com/pdf/pharmaceutical-excipients-a-review-44zn8puj1t.pdf

[15] Ku MS. Use of the biopharmaceutical classification system in early drug development. AAPS J 2008; 10(1): 208-12.
[http://dx.doi.org/10.1208/s12248-008-9020-0] [PMID: 18446521]

[16] Verma P, Thakur A, Deshmukh K, Jha A, Verma S. Research Routes of drug administration. Int J Pharm Stud Res 2010; 1: 54-9. Available from: https://romanpub.com/resources/ijpsr%20v11-202--7.pdf

[17] Parashar SK, Srivastava SK, Dutta NN, Garlapati VK. Engineering aspects of immobilized lipases on esterification: A special emphasis of crowding, confinement and diffusion effects. Eng Life Sci 2018; 18(5): 308-16.
[http://dx.doi.org/10.1002/elsc.201700082] [PMID: 32624910]

[18] Surendra KP, Sunil KS, Vijay KG. The mathematical modeling for the optimization of triacyl glycerol acylhydrolase production through artificial neural network and genetic algorithm. Int J Pharma Bio Sci 2019; 10(3): 135-43.
[http://dx.doi.org/10.22376/ijpbs.2019.10.3.b135-143]

[19] Srivastava SK. New challenges on natural resources and their impact on climate change in the indian context. In: Islam M.N., van Amstel A. (eds) India: Climate Change Impacts, Mitigation, and Adaptation in Developing Countries. Springer Climate. Springer, Cham. 2021; 1:1-15.
[http://dx.doi.org/10.1007/978-3-030-67865-4_1]

[20] Parashar SK, Srivastava SK, Garlapati VK, Dutta NN. Production of microbial enzyme triacylglycerol Acylhydrolases by ASPERGILLUS SYDOWII JPG01 in submerged fermentation using agro residues. Asian J Microbiol Biotechnol Environ Sci 2019; 21(4): 1076-9. Available from: http://www.envirobiotechjournals.com/article_abstract.php?aid=10217&iid=295&jid=1

[21] Srivastava SK, Islam MN. Review on Carbon Footprint in India and South East Asia in Comparison to the World Average in the Last Few Centuries. In: Islam MN, Amstel Av, Eds. India III: Climate Change and Landscape Issues in India Springer Climate.
[http://dx.doi.org/10.1007/978-3-031-85126-1_17]

[22] Srivastava SK, Dhaker KL. Data-driven approach for synthesizing, characterization, and experimental investigation of biomass-based plastic. Next Materials 2025; 8: 100596.
[http://dx.doi.org/10.1016/j.nxmate.2025.100596]

[23] Cardoso RV, Pereira PR, Freitas CS, Paschoalin VMF. Trends in drug delivery systems for natural bioactive molecules to treat health disorders: The importance of nano-liposomes. Pharmaceutics 2022; 14(12): 2808.
[http://dx.doi.org/10.3390/pharmaceutics14122808] [PMID: 36559301]

[24] Diebold Y, Calonge M. Applications of nanoparticles in ophthalmology. Prog Retin Eye Res 2010; 29(6): 596-609.
[http://dx.doi.org/10.1016/j.preteyeres.2010.08.002] [PMID: 20826225]

[25] Levy-Nissenbaum E, Radovic-Moreno AF, Wang AZ, Langer R, Farokhzad OC. Nanotechnology and aptamers: applications in drug delivery. Trends Biotechnol 2008; 26(8): 442-9.
[http://dx.doi.org/10.1016/j.tibtech.2008.04.006] [PMID: 18571753]

[26] Huxford RC, Della Rocca J, Lin W. Metal–organic frameworks as potential drug carriers. Curr Opin Chem Biol 2010; 14(2): 262-8.
[http://dx.doi.org/10.1016/j.cbpa.2009.12.012] [PMID: 20071210]

[27] Horcajada P, Serre C, Vallet-Regí M, Sebban M, Taulelle F, Férey G. Metal-organic frameworks as

efficient materials for drug delivery. Angew Chem Int Ed 2006; 45(36): 5974-8. [http://dx.doi.org/10.1002/anie.200601878] [PMID: 16897793]

[28] Rieter WJ, Taylor-Pashow K, An H, Lin W. Nanoscale metal-organic frameworks as potential multimodal contrast enhancing agents. J Am Chem Soc 2006; 128(28): 9024-5. [http://dx.doi.org/10.1021/ja0627444]

[29] Srivastava SK, Dhaker KL. Data-driven approach for Cu recovery from hazardous e-waste. Process Saf Environ Prot 2024; 183: 665-75. [http://dx.doi.org/10.1016/j.psep.2024.01.013]

[30] Garlapati VK, Parashar SK, Klykov S, *et al.* Invasive weed optimization coupled biomass and product dynamics of tuning soybean husk towards lipolytic enzyme. Bioresour Technol 2022; 344(Pt B): 126254. [http://dx.doi.org/10.1016/j.biortech.2021.126254] [PMID: 34757227]

[31] Langer R. Drug delivery and targeting. Nature 1998; 392(6679) (Suppl.): 5-10. Available from: https://pubmed.ncbi.nlm.nih.gov/9579855/ [PMID: 9579855]

[32] Adepu S, Ramakrishna S. Controlled drug delivery systems: Current status and future directions. Molecules 2021; 26(19): 5905. [http://dx.doi.org/10.3390/molecules26195905] [PMID: 34641447]

[33] Srivastava SK. Advancement in biogas production from the solid waste by optimizing the anaerobic digestion. Waste Dispos Sustain Energy 2020; 2(2): 85-103. [http://dx.doi.org/10.1007/s42768-020-00036-x]

[34] Langer R. New methods of drug delivery. Science 1990; 249(4976): 1527-33. [http://dx.doi.org/10.1126/science.2218494] [PMID: 2218494]

[35] Shargel L, Andrew B, Wu-Pong S. Applied biopharmaceutics & pharmacokinetics. Appleton & Lange Stamford; Stamford, UK 1999;264. Available from: https://akfarstfransiskusxaverius.ac.id/wp-content/uploads/2023/08/3_Applied-Biopharmaceutics-Pharmacokinetics-PDFDrive-.pdf

[36] Srivastava SK. Assessment of groundwater quality for irrigation suitability and its impacts on crop yields in the Guna district, India. Agric Water Manage 2019; 216: 227. [http://dx.doi.org/10.1016/j.agwat.2019.02.005]

[37] Jambhekar SS, Breen PJ. Basic Pharmacokinetics. Pharmaceutical Press; London, UK: 2009;76. Available from: https://www.pharmaceuticalpress.com/product/basic-pharmacokinetics-third-edit-on-copy/

[38] Fleisher D, Li C, Zhou Y, Pao L, Karim A. Drug, meal and formulation interactions influencing drug absorption after oral administration. Clinical implications. Clin Pharmacokinet 1999; 36(3): 233-54. [http://dx.doi.org/10.2165/00003088-199936030-00004] [PMID: 10223170]

[39] Seydel JK, Wiese M. Drug-membrane interactions: Analysis, drug distribution, modeling. John Wiley & Sons; Hoboken, NJ, USA 2009;15. ISBN: 978-3-527-61649-7. Available from: https://www.wiley.com/en-us/Dru--Membrane+Interactions%3A+Analysis%2C+Drug+Distribution%2C+Modeling-p-9783527616497

[40] Gillette JR. Factors affecting drug metabolism. Ann N Y Acad Sci 1971; 179(1): 43-66. [http://dx.doi.org/10.1111/j.1749-6632.1971.tb46890.x] [PMID: 4398309]

[41] Ekins S, Ring BJ, Grace J, McRobie-Belle DJ, Wrighton SA. Present and future *in vitro* approaches for drug metabolism. J Pharmacol Toxicol Methods 2000; 44(1): 313-24. [http://dx.doi.org/10.1016/S1056-8719(00)00110-6] [PMID: 11274898]

[42] Taft DR. Pharmacology. Elsevier; Amsterdam, The Netherlands: Drug excretion 2009; 175–199. [http://dx.doi.org/10.1016/B978-0-12-369521-5.00009-9]

[43] Hallare J, Gerriets V. Half Life. Stat Pearls Publishing LLC; Treasure Island, FL, USA: 2020.

Available from: https://solutionsretreat.org/wp-content/uploads/nih-half-life.pdf

[44] Srivastava SK, Ramanathan AL. Geochemical assessment of groundwater quality in vicinity of Bhalswa landfill, Delhi, India, using graphical and multivariate statistical methods. Environ Geol (Berl) 2008; 53(7): 1509-28.
[http://dx.doi.org/10.1007/s00254-007-0762-2]

[45] Srivastava SK, Ramanathan AL. Assessment of landfills vulnerability on the groundwater quality located near floodplain of the perennial river and simulation of contaminant transport. Model Earth Syst Environ 2018; 4(2): 729-52.
[http://dx.doi.org/10.1007/s40808-018-0464-7]

[46] Srivastava SK, Ramanathan AL. Geochemical assessment of fluoride enrichment and nitrate contamination in groundwater in hard rock aquifer using graphical and statistical methods. Journal of Earth System Sciences2018b;127:104 (1-23).
[http://dx.doi.org/10.1007/s12040-018-1006-4]

[47] Prabu SL, Suriyaprakash T, Ruckmani K, Thirumurugan R. Basic pharmacokinetic concepts and some clinical applications Intech Open. London, UK: Biopharmaceutics and Pharmacokinetics 2015.
[http://dx.doi.org/10.5772/58502]

[48] Rieter WJ, Pott KM, Taylor KML, Lin W. Nanoscale coordination polymers for platinum-based anticancer drug delivery. J Am Chem Soc 2008; 130(35): 11584-5.
[http://dx.doi.org/10.1021/ja803383k] [PMID: 18686947]

[49] Khoz R, Yazdian F, Pourmadadi M, Rahdar A, Fathi-karkan S, Pandey S. Current trends in silica based drug delivery systems. Eur J Med Chem Rep 2024; 12: 100206.
[http://dx.doi.org/10.1016/j.ejmcr.2024.100206]

[50] Fang M, Liu R, Fang Y, Zhang D, Kong B. Emerging platelet-based drug delivery systems. Biomed Pharmacother 2024; 177: 117131.
[http://dx.doi.org/10.1016/j.biopha.2024.117131] [PMID: 39013224]

[51] Guo YX, He YX. Nanoparticle-based drug delivery systems: An updated strategy for treating fungal keratitis. Colloid Interface Sci Commun 2024; 61: 100794.
[http://dx.doi.org/10.1016/j.colcom.2024.100794]

[52] Ye B, Xiang R, Luo F. Hydrogel-Based drug delivery systems for diabetes bone defects. Chem Eng J 2024; 497: 154436.
[http://dx.doi.org/10.1016/j.cej.2024.154436]

[53] Srivastava SK. Soil and Environmental variables influencing Green House Gas cycling in an agroecosystem. Book: Greenhouse Gas Regulating Microorganisms in Soil Ecosystems - Perspectives for Climate Smart Agriculture 2024; Chap 2, page 31-42. ISBN 3031705688, 9783031705687.
[http://dx.doi.org/10.1007/978-3-031-70569-4_2]

[54] Srivastava SK. Chemical speciation and leaching of trace metals in groundwater from the depleted landfills, India. J Geol Soc India 2023; 99(4): 554-62.
[http://dx.doi.org/10.1007/s12594-023-2344-7]

[55] Srivastava SK, Shrivastava R. Modeling for Copper Recovery from E-Waste by Using Machine Learning Technique: An Approach for the Circular Economy. Curr Anal Chem 2024.
[http://dx.doi.org/10.2174/0115734110343714241130170537]

[56] Srivastava SK. Advance mathematical modeling for the delineation of the groundwater potential zone in Guna district, India. Model Earth Syst Environ 2023; 9(2): 1793-809.
[http://dx.doi.org/10.1007/s40808-022-01587-0]

[57] Srivastava SK. Geochemical assessment of fluoride enrichment and nitrate contamination in groundwater in a hard rock aquifer by using graphical and statistical methods. Journal of Earth System Science 2018; 127 (7): 104 (1-23).
[http://dx.doi.org/10.1007/s12040-018-1006-4]

[58] Srivastava SK, Gautam AK, Singh P. Statistical evaluation of Recovery of copper from e-waste by using the Hydrometallurgical Method and RSM model" Journal of Environmental Science, Toxicology and Food Technology" Oct. 2016; 10 (10) (ver. II):31-43. Available from: www.iosrjournals.org/iosr-jestft/papers/vol10-issue10/Version-2/F1010023143.pdf

CHAPTER 2

Antidiabetic Properties of Metal Complexes: Recent Advancements

Divyakshi Arya[1], **Gulshan A. Khan**[1], **Shweta Singh**[1], **Anjali Upadhyay**[1], **M.L. Sharma**[2], **Ashish R. Dwivedi**[3], **Som S. Dubey**[1] and **Sarvesh K. Pandey**[1,*]

[1] *Department of Chemistry, D.D.U. Gorakhpur University, Gorakhpur, Uttar Pradesh, India*

[2] *Central Department of Chemistry, Tribhuvan University, Kirtipur, Kathmandu, Nepal*

[3] *GITAM School of Pharmacy, GITAM University, Hyderabad, India*

Abstract: The demographic count of individuals diagnosed with diabetes has risen worldwide over the past 10 years. Today, a majority of countries that are experiencing a high prevalence of diabetes are developing countries. Due to the concerning prevalence of diabetes, chemists and pharmacists are always looking for and creating novel, powerful medications to treat the condition. The metallo-drug interaction mechanism is receiving a lot of interest these days. In medicine, metals and organometallic complexes are utilized to treat a variety of illnesses. Pronounced advancements have been made in the deployment of coordination compounds of transition metals as medications to alleviate numerous human ailments. Transition metals may engage with a variety of anionic molecules and exhibit a range of oxidation states. The present status and probable upcoming ways of metal complexes of V(IV), Zn(II), Cr(III), Cu(II), and Co(II) as possible oral drug candidates against type 2 diabetes have recently reviewed. This chapter summarizes key developments in the synthesis and therapeutic assessment of metal complexes for diabetes management.

Keywords: Diabetes, Insulin mimetic property, Metal complexes, Transition metals.

INTRODUCTION

Diabetes is a chronic metabolic imbalance triggered by either inadequate insulin secretion by the pancreatic tissue or the body's reduced sensitivity to insulin, impairing glucose regulation. Insulin is a hormone responsible for modulating blood sugar levels. If not regulated and managed, diabetes gradually leads to hyperglycemia, high blood sugar, which eventually damages very sensitive and

* **Corresponding author Sarvesh K. Pandey:** Department of Chemistry, D.D.U. Gorakhpur University, Gorakhpur, Uttar Pradesh, India; E-mail: skpandey1982@yahoo.co.in

Abhay Nanda Srivastva (Ed.)
All rights reserved-© 2026 Bentham Science Publishers

delicate systems in the body, notably nerves and blood vessels. There are different categories of diabetes mellitus that have been described (Fig. **1**).

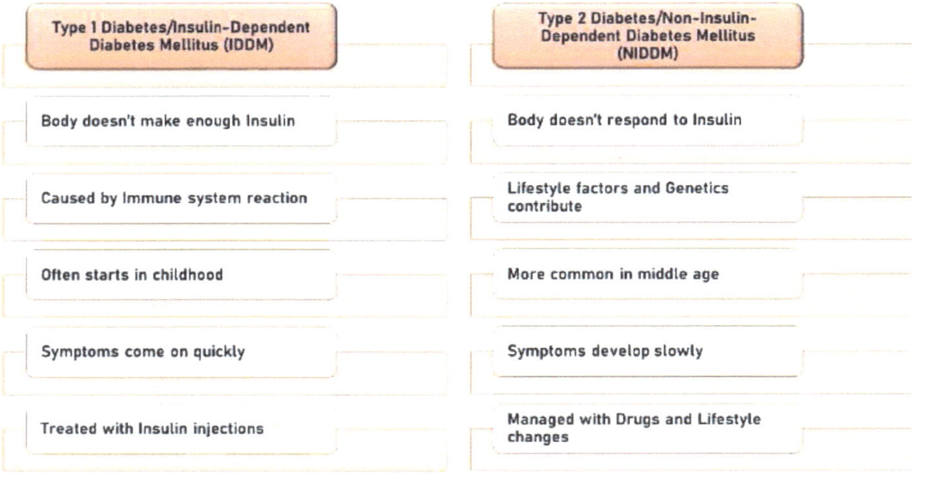

Fig. (1). Types of diabetes.

As of 2022, diabetes had impacted about 14% of adults aged 18 and older, approximately doubling since 1990. Among adults aged 30 and older with diabetes, 59% were not given any treatment to manage diabetes, and coverage for diabetes treatment was exceptionally low in low- and middle-income countries.

In 2021, diabetes was directly liable for 1.6 million deaths, roughly half of them (47%) occurring prior to the age of 70. Likewise, diabetes caused another 530,000 deaths from kidney disease and resulted in about 11% of cardiovascular deaths due to high blood glucose levels [1].

Global Prevalence

An estimated 240 million people worldwide live with unidentified diabetes, of whom approximately half of the adults affected are oblivious to having diabetes. Misdiagnosis, accompanied by the rising incidence of diabetes, places a hefty burden on the healthcare systems globally. At present, 537 million individuals, ranging in age from 20 to 79 years, with diabetes, impact nearly 10.5% of the global population. The International Diabetes Federation (IDF) determined that the healthcare expenses associated with diabetes reached $966 billion in 2021, which is anticipated to escalate to more than $1,054 billion in 2045. Consequently, the prevalence of diabetes is foreseen to escalate sharply to 643 million (11.3%) people in 2030 and 783 million (12.2%) in 2045 (Fig. **2**). This calls for escalating global efforts to prevent, diagnose early, and manage diabetes [2].

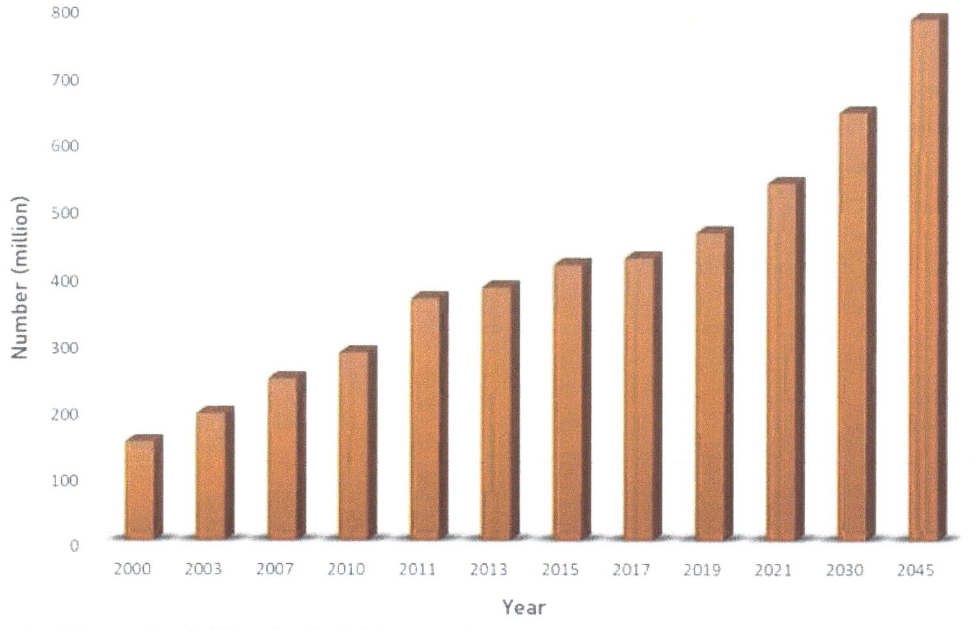

Fig. (2). The number (millions) of individuals aged 20-79 years having diabetes across the globe (envisioned numbers in case of 2030 and 2045).

The top 10 countries with the elevated incidence in the world are India, China, the USA, Pakistan, Russia, Japan, Indonesia, Bangladesh, Brazil, and Italy. The escalating health condition is now becoming a global challenge. Prevalence of diabetes consistently is amplifying four times faster in low- and middle-income nations than in the high-income nations [3]. The diabetic community is primarily found in low- or middle-income countries, and it corresponds to nearly 80% of the global dispersal, hence protruding an uneven spread of the disease [4].

Associated Risk Factors

Diabetes Mellitus (DM) is highly pervasive in populations that are swiftly undergoing transitions from a traditional to a modern lifestyle, which confers an expanded risk of the ailment. Type 2 diabetes is strongly correlated with overweight or obesity, older age, certain ethnicities, and a genetic linkage to diabetes within the family, although type 2 diabetes has an unknown etiology [5]. The evolution results due to the interaction of complex genetic (polygenic) and environmental factors [4].

The risk for developing diabetes mellitus is not only influenced by genetic predispositions but also by numerous other factors. The worldwide frequency of type 2 diabetes is expanding, and it is contemplated that population aging, expansion of the middle class, and rise in urban development are contributing factors for growing rates of obesity [5]. The most remarkable risk factor for diabetes is obesity, with an approximation of nearly 60% of type 2 cases [6], and it remains the primary reason for Disability Adjusted Life Years (DALYs) ascribed to diabetes in regions like North Africa and the Middle East, Latin America and the Caribbean, and Central and Eastern Europe [2].

Additional contributing factors are high anthropometric index related to body mass/BMI (52.2%), poor nutritional choices (25.7%), nature-related and workplace stress (19.6%), tobacco use (12.1%), lack of bodily movement (7.4%), and consumption of alcoholic beverages (1.8%) [7]. Significantly, the arrival of diabetes can often be delayed through suitable diet and lifestyle modifications.

Besides acute complications, there are significant long-term complications of diabetes, comprising cardiovascular and kidney diseases that significantly contribute to the elevated risk of drastic outcomes and mortality among hospitalized individuals with diabetes. These complications underline the necessity for early intervention and comprehensive management strategies [2].

Tackling Strategies

Type 2 Diabetes Mellitus (DM) can generally be checked or deferred through drug-free approaches such as regular physical activity, reduced sedentary behavior, quitting smoking, and maintenance of a healthy weight. Weight loss is very effective, as loss of each kilogram reduces the relative risk of diabetes by 16% and greatly reduces ailment progression [8]. An arbitration aimed at diet, exercise, or a fusion of diet and exercise diminished the inclination of developing diabetes by 31%, 46%, and 42%, respectively, in a clinical study performed in Da Qing, China [9]. Such findings underscore the importance of lifestyle modifications in the prevention of diabetes.

An individual at an expanded risk of having diabetes should undergo repeated screenings. Risk can be evaluated based on personal risk factors (including older age, membership of a marginalized racial group, and ancestral background of diabetes), clinical signs (such as hypertension, elevated cholesterol, and excess body weight), and prior occurrences of glucose abnormalities (like diabetes gravidarum, early-stage glucose intolerance, or glucose regulation impairment). The keys that reduce the impact of diabetes are early detection and proactive management [10].

METAL COMPLEXES AND THEIR ANTI-DIABETIC PROPERTIES

A metal complex, or coordination complex, is a chemical species formed by a central atom/ion, generally a metal referred to as the coordination center, which encircles a group of molecules/ions known as ligands or complexing agents [11]. Numerous compounds consisting of metal, particularly those comprising transition metals (elements from the d-block of the periodic table, such as vanadium, zinc, *etc.*), are coordination complexes of metal complexes [12a]. Pronounced advances have been made in harnessing transition-metal complexes as therapeutics to mitigate numerous human ailments. D-block elements are capable of binding with numerous anionic molecules and demonstrating diverse oxidation states. Their chemical versatility has contributed to the advancement of metal-based therapeutics, presenting revitalizing pharmacological applications and emerging treatment potentials[13]. Here, in this chapter, we will discuss some metal complexes and their anti-diabetic properties:

a. **Vanadium:** The chemical behaviour of vanadium is particularly remarkable for its access to four successive oxidation states, 2 to 5; in contrast, niobium and tantalum, other members of group 5, primarily exhibit chemistry in the +5 oxidation state [12a]. Vanadium(II) and Vanadium(III) complexes are, in general, relatively exchange-inert and show reducing properties; conversely, Vanadium(IV) and Vanadium(V) complexes act as oxidants (Fig. **3**). Due to the vanadium ion's size being quite large in comparison, some of its complexes can reach coordination no. larger than 6, for instance $[V(CN)_7]^{4-}$.

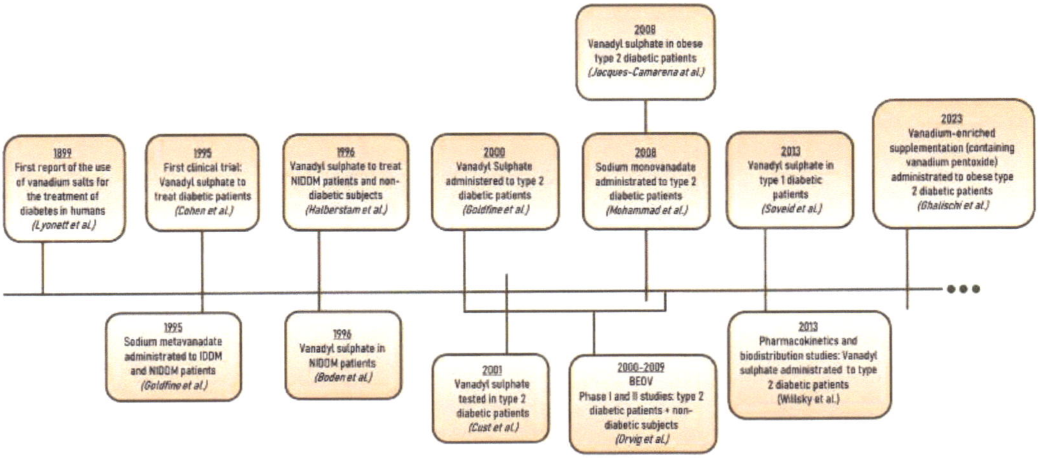

Fig. (3). The recent relevance and chronology of key human studies exploring the therapeutic application of vanadium compounds in diabetes treatment[12b].

In 2015, Debbie C. Crans studied the anti-diabetic properties of organic vanadate esters. The compounds were reported based on the stance that the underlying chemistry and characteristics of vanadium-based organic ester compounds influence their effect as phosphatase suppressors, informed by framework-related insights derived from complexes involving vanadium and phosphatase (Fig. **4**) [14]. Organic vanadate esters are classically four-coordinate (substrate analogues) or five-coordinate (transition-state analogue inhibitors). Phosphatases, a distinct group of often membrane-associated enzymes, perform based on the substrate and cellular site. Whereas phosphate hydrolysis typically follows dissociative or associative pathways, PTP1B operates *via* an intensive exploded mechanism, with elongated vanadium-oxygen bonds imitating transition-state bond breaking. Gaining insight into the insulin signaling pathway is equally important (Fig. **5**).

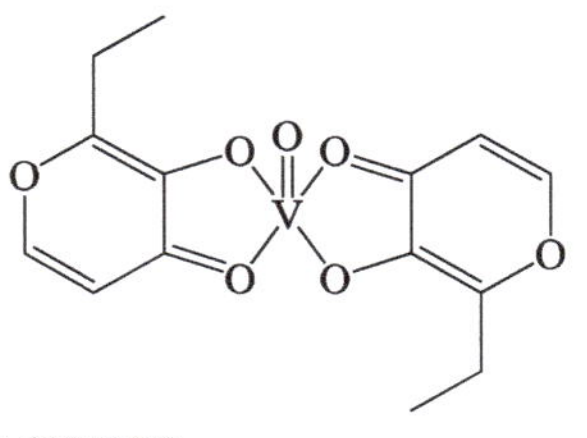

Fig. (4). Bis(ethylato)oxovanadium(IV) (BEOV).

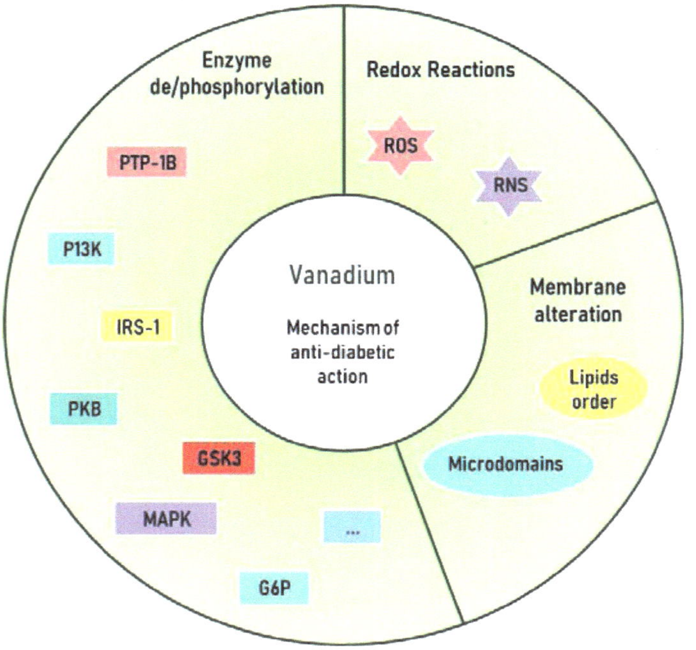

Fig. (5). Illustrative scheme depicting how vanadium compounds exert their anti-diabetic activity, including: **(1)** Enzyme de/phosphorylation, **(2)** Membrane alteration, and **(3)** Redox reactions [12b].

In 2016, Mohini *et al.* reported on the diabetes-modulating potential of the vanadium complex of the isolated thespesone from the bark of *Thespesia populnea* (Malvaceae). The vanadium complex, administered orally (5-20 mg/kg) to diabetic mice for 22 days, significantly lessens plasma glucose and triglycerides while increasing body weight. It enhanced insulin sensitivity, prevented dexamethasone-induced metabolic changes, and boosted glucose uptake, suggesting potential for treating Type II diabetes. It is useful due to its ability to lessen insulin resistance [15].

In 2017, Tamás Jakusch and Tamás Kiss studied *in-vivo* about the insulin-analogous functions of vanadium complexes. McNeill *et al.* pioneered vanadium's insulin-mimetic effects *in vivo* by supplementing the drinking water of Streptozotocin (STZ)-induced diabetic rats with sodium orthovanadate over a period of several weeks. Afterwards, $VO(IV)SO_4$ is used instead of the only partly soluble vanadates, and the outcome was a couple of negative side effects. Orally administered vanadium complexes undergo GI absorption and serum transport, likely undergoing biotransformation that likely substitutes the original ligand with plasma components before cellular uptake. The active species is alleged to be vanadate ($H_2VO_4^-$), which may act by inhibiting protein tyrosine phosphatases or activating tyrosine kinases in insulin signaling. Furthermore, redox processes and ROS generation may contribute to the antidiabetic effects, although encouraging animal results have not yet advanced to human clinical trials (Fig. **6**) [16].

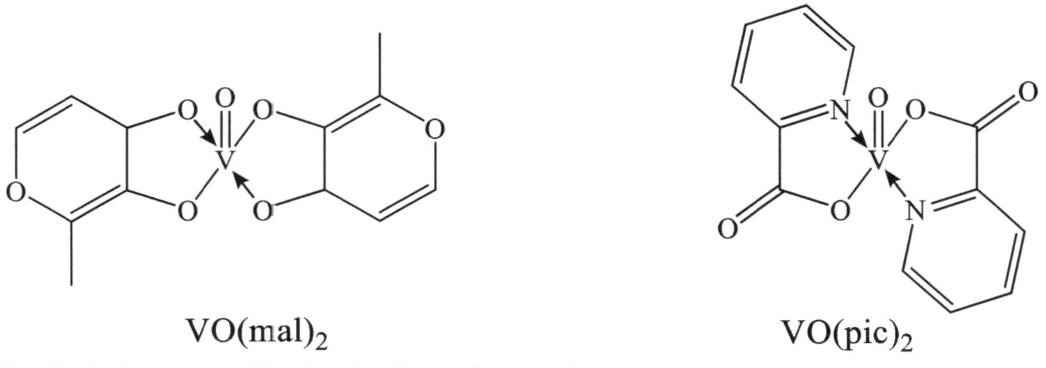

Fig. (6). Basic structure of insulin-mimetic vanadium complexes.

In 2019, N. Patel *et al.* studied vanadium complexes formed by the interaction of NNO donor Schiff base ligands with vanadyl sulphate. Their anti-diabetic potential, demonstrated through α-amylase and α-glucosidase inhibition assays, highlights their promise as effective insulin-enzyme inhibitors (Fig. **7**) [17].

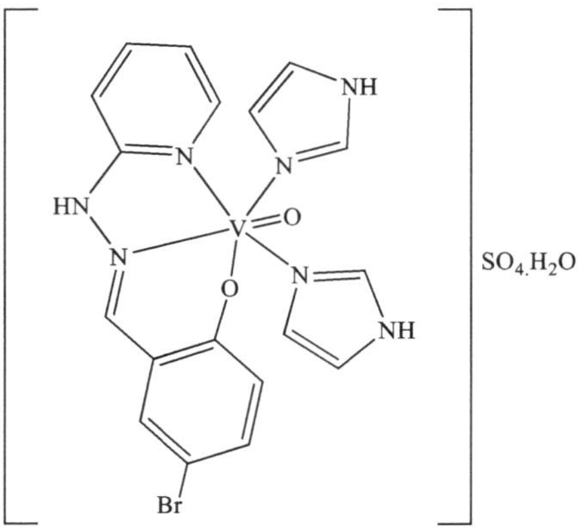

Fig. (7). Oxovanadium complex of Schiff base of 2-hyrazinopyridine and 5-bromosalicylaldehyde.

In 2020, Neetu Patel and colleagues conducted studies on two new dioxidovanadium(V) complexes comprising the Nicotinic acid (2-hydroxy-3-methoxy-benzylidene)-hydrazide (H2L) and an imidazole-based tridentate ligand. The α-amylase and α-glucosidase inhibitory activities exhibited by these species highlight their potential as promising inhibitors of insulin-related enzymes in anti-diabetic therapy. The complexes displayed noteworthy α-glucosidase inhibition in a concentration-dependent way, signifying dose-dependent activity of the oxidovanadium complexes. Correspondingly, *in vitro* α-amylase inhibition was detected for both ligands and complexes, compared with acarbose, and stated as IC_{50} values and percentage inhibition. All complexes demonstrated increasing inhibition with higher concentrations (Fig. **8**) [18].

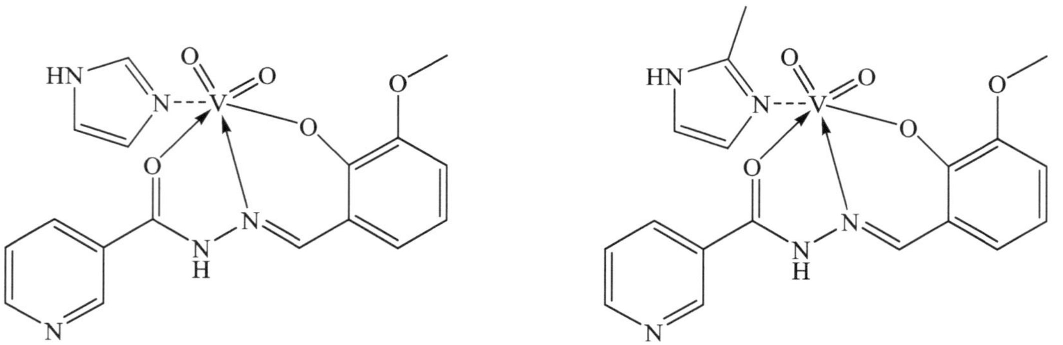

Fig. (8). Dioxidovanadium (V) complexes $[VO_2(L)]$Him and $[VO_2(L)]$H2Meim (Him = Imidazole, H2Meim = methylimidazole and H_2L = Nicotinic acid (2-hydroxy-3-methoxy-benzylidene)- hydrazide ONO donor ligand).

In 2021, Sabina Semiz studied vanadium complexes as a possible therapeutic agent due to their antihyperglycemic effects. Vanadium compounds could be classified as adjuncts to standard COVID-19 therapy, given the favorable anti-hyperglycemic and anti-inflammatory effects and the possible mechanistic link between diabetes and COVID-19. BMOV and its ethylmaltol correspondent, bis(ethylmaltolato)oxovanadium(IV) (BEOV), displayed analogous antidiabetic effects in Zucker Fatty and STZ-diabetic Wistar rats. BEOV, with enhanced pro-drug features, advanced to phase 1 and 2 clinical trials. In T2D patients, it reduced fasting glucose, HbA1c levels, and refined oral glucose intolerance, while other trials in T1D and T2D subjects exhibited lower insulin requirements, designatingescalated insulin sensitivity. In particular, vanadium treatment did not induce hypoglycemia and had no effect on C-peptide levels in T1D patients, indicating its action was independent of insulin secretion (Fig. **9**) [19].

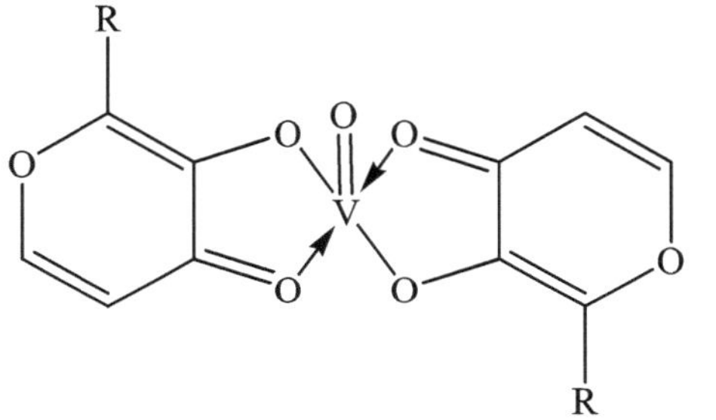

Fig. (9). Bis(maltolato)oxovanadium(IV) (BMOV), R = CH_3, and bis(ethylmaltolato)oxovanadium(IV) (BEOV), R = C_2H_5.

In 2023, Jiayue Xia *et al.* studied that among various micronutrients, supplements including vitamin E, vitamin K, vanadium, chromium, and niacin have been found to offer enhanced benefits for the treatment of type 2 diabetes mellitus. Anomalous lipid profiles are common in T2DM and increase cardiovascular risk. Vanadium supplementation exhibited the sturdiest cholesterol-lowering effect, with vanadium mainly stored in bones, liver, and kidneys. Vanadium compounds, including vanadyl sulfate and sodium metavanadate, lower plasma cholesterol without altering triglycerides or free fatty acids, likely by inhibiting cholesterol synthesis. In hepatocytes, sodium metavanadate stimulates lipogenesis and suppresses lipolysis, thereby regulating lipid metabolism. While favorable in diabetes, hyperlipidemia, and hypertension, vanadium supplementation is pointless for the overall population, and dosage must be prudently controlled due to potential toxicity [20].

In 2024, Anna Jurowska *et al.* reported work on an array of dioxidovanadium(V)-based ionic complexes featuring Schiff base ligands derived from 2-hydroxybenzhydrazide and substituent-bearing salicyladehydes. The results were compared with previously studied electrically balanced complexes of vanadium bearing similar Schiff base ligands, investigated earlier as possible insulin-resembling agents. Several compounds with varying substituents on the aromatic rings of salicylaldehyde- and benzhydrazide-derived fragments were tested. Many complexes exhibited higher biological activity than known insulin-mimetic compounds such as $VOSO_4$ and BMOV. However, limitations included difficulties in structural determination and poor water solubility (Fig. **10**) [21].

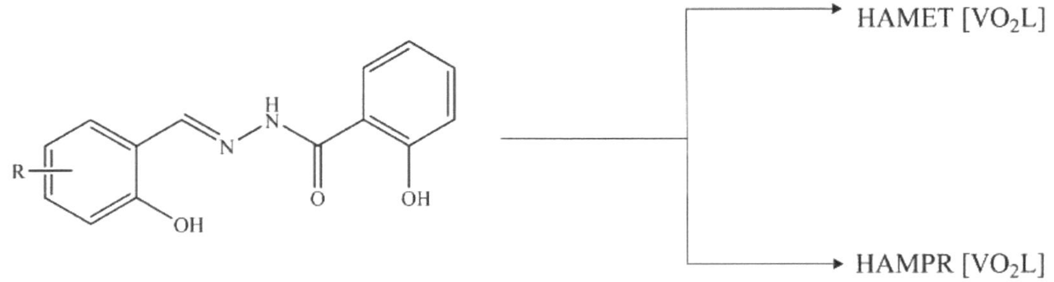

Fig. (10). Dioxido complexes of vanadium (V) with Schiff base-type having the cation, protonated 2-aminoethanol (HAMET) or 3-amino-1-propanol (HAMPR).

b. **Zinc:** Usually, most of its compounds comprise zinc with an oxidation state of +2. Zinc can be categorized as a post-transition element within the main group. Unlike other compounds in the same oxidation state, Zinc(II) compounds are distinguished by a lack of vivid coloration, structural symmetry, and minimal participation in redox reactions [22].

In 2017, Ryota Saito *et al.* had formed a newly designed molecule 1-(arylmethyl)-2,5-dihydro-4-hydroxy-5-oxo-1H-pyrrole-3-carboxamides. The compound was fabricated *via* N-tritylated acrylamide as the precursor, and its Zn(II) complex (Fig. **11**) was readily formed by mixing it with $ZnSO_4$ in the presence of LiOH. The research highlights the hypoglycemic potential of newly synthesized Zinc(II) complexes with N2O2 coordination derived from pyrrole-3-carboxamide. To evaluate the insulin-mimetic activity of the fabricated zinc(II) complexes *in vitro*, their inhibitory effects on Free Fatty Acid (FFA) release from epinephrine-stimulated rat adipose cells were assessed. This facile and consistent method exposed that all complexes suppressed FFA release in a dose-dependent manner, authorizing their insulin-mimetic potential [23].

In 2018, Franc Perdih *et al.* reported that some zinc complexes also exhibit insulin-mimetic activity. The two different spatial arrangements are shown in the Zn(II) complexes: cis and trans, with octahedral geometry. Zinc bis(3-hydroxypicolinato) complex that is $[Zn(Hhypic)_2(H_2O)_2]$ is more active than the other two complexes, [zinc bis(3,5-difluoropicolinato)] complex and [zinc bis(4-(hydroxymethyl)picolinato)] complex. The insulin-mimetic activity of the designated complexes was assessed by their ability to inhibit epinephrine-induced

Free Fatty Acid (FFA) release from isolated rat adipocytes. All assessed metal complexes displayed insulin-mimeticactivity. Among them, zinc complexes were active, with the Hhypic complex being the most effective; however, their activity was not higher than that of $ZnSO_4$(Fig. **12**) [24].

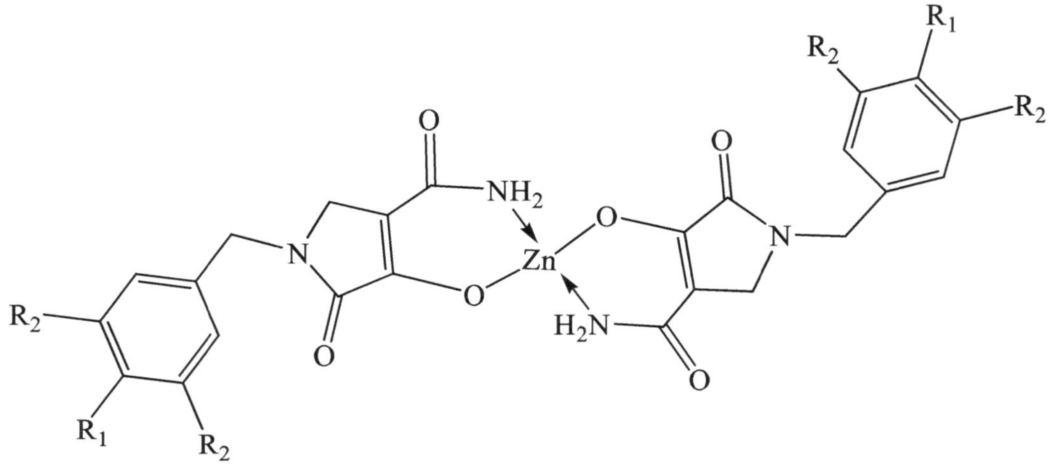

Fig. (11). Zinc(II) complex of (arylmethyl)-2,5-dihydro-4-hydroxy-5-oxo-1H-pyrrole-3-carboxamides.

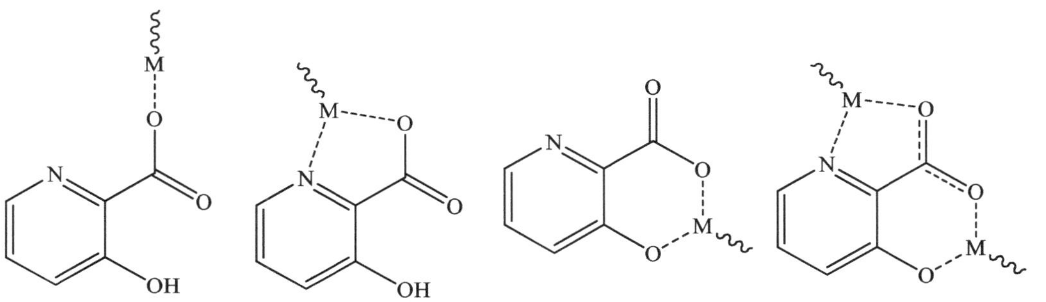

M(Hypic-O)	**M(Hhypic-N,O)**	**M(Hhypic-O,O)**	**M_2(hypic-N,O,O',O")**
1	**2**	**3**	**4**

Fig. (12). Probable binding modes of H_2hypic. **(i)** Monodentate Hhypic - O, **(ii)** Bidentate Hhypic-N, O, **(iii)** Bidentate Hhypic - O, O, and **(iv)** Bridging hypic - N, O, O', O" ligation modes.

In 2020, Nagaraj Govindan *et al.* produced eco-compatible Zinc-infused Catharanthus roseus (Sadabahar or Baramasi) plants *via* a labile photon-assisted approach for the development of green biomaterials. Phytochemicals present in the extracted leaf solution of *C. roseus* contribute to the formation of nanoparticles by acting as natural reducing agents, making them suitable for improved antidiabetic applications.The Zn-doped *C. roseus* extract revealed strong α-amylase inhibitory activity with a satisfactory IC_{50} value. Assimilation of Zn enhanced blood sugar regulation, leading to efficient suppression of fasting glucose levels, likely due to amino acids present in the nanoparticles. These biosynthesized Zn-doped *C. roseus* NPs thus hold promise as safe, biocompatible starch blockers for diabetes management, offering a benefit over synthetic medicines that often cause side effects such as bloating and diarrhea [25].

In 2020, Ginés M. Esteban-Parra *et al.* synthesized the newly developed 7-amino-5-methyl-1,2,4-triazolo[1,5-a]pyrimidine ligand and an emerging group of zinc-based coordination complexes by X-ray crystallography. Compound 1 $[Zn(7\text{-amtp})_2(H_2O)_4]\cdot(NO_3)_2\cdot2(7\text{-amtp})\cdot6H_2O$ and compound 2 $[Zn(7\text{-amtp})_2(H_2O)_4]SO_4\cdot 1\cdot5H_2O$ a mononuclear complexes in which 7-amine-5-methyl-1,2,4-triazol-[1,5-a]pyrimidine coordinates by N_3 atom to zinc ions. Half an hour following glucose intake, compound 1 exhibits a promising glycemic-lowering effect and effectively regulates glycemic concentration in physiologically intact mice. Although differently, compound 2 is unable to stabilize glycemic levels upon test completion, but does inhibit the induced response, leading to only a 13% increase in glycemia (Fig. **13**) [26].

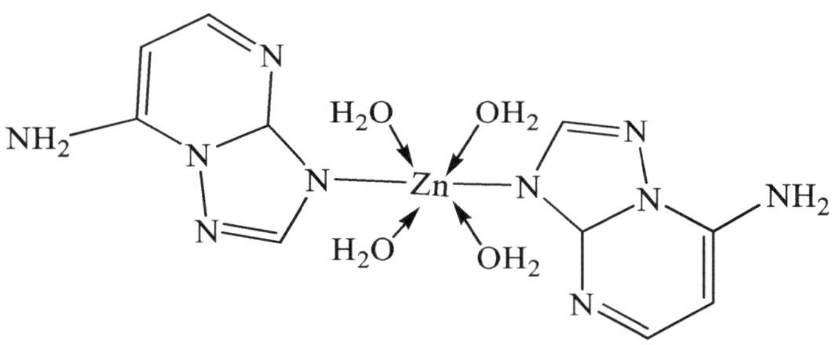

Fig. (13). Zinc(II) complex of (7-amino-5-methyl[1, 2, 4]triazolo[1,5-a]pyrimidine.

c. **Chromium:** The group 6 transition metal, Chromium generally acquires the oxidation states +3 and +6. Although +2 is less prevalent, these three oxidation states are quite common. Far less common are the oxidation states +1, +4, and +5 [12].

In 2015, M. A. Mahmoud *et al.* worked on a series of Cr3+ complexes *via* a template reaction using metformin-based Schiff bases with various aldehydes. Notably, $[CrL^4Cl(H_2O)_2].3H_2O$ and $[CrL^5Cl(H_2O)_2].2½H_2O$ exhibited significant hypoglycemic activity. In this analysis, alloxan (150 mg/kg) prompted hyperglycemia in mice, with fasting blood glucose >200mg/dl, confirming diabetes. After 14 days of use, metformin significantly lowered blood glucose levels compared with the diabetic control group. Likewise, chromium complexes $[CrL^4Cl(H_2O)_2].3H_2O$ and$[CrL^5Cl(H_2O)_2].2½H_2O$ exhibited noticeable antihyperglycemic effects. At 20 mg/kg, they mitigated glucose levels by 4.24% and 24.62%, although at 40mg/kg, reductions extended 65.74% and 66.76%, respectively. Survival rates (83.33-100%) were comparable to controls, signifying decent safety. These conclusions highlight the favorable blood-glucose-lowering potential of chromium complexes (Fig. **14**) [27].

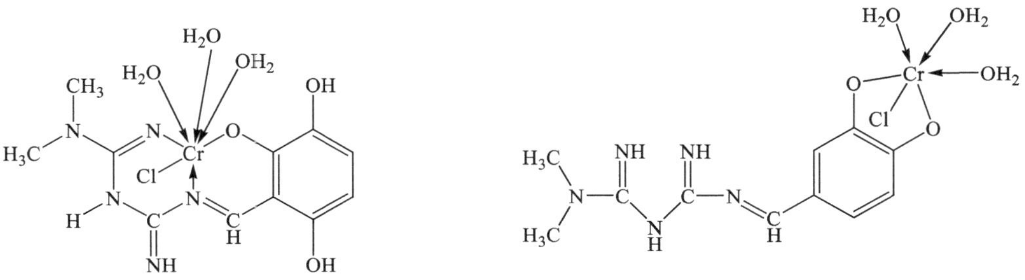

Fig. (14). $[CrL^4Cl(H_2O)_2].3H_2O$ and $[CrL^5Cl(H_2O)_3].2½H_2O$.

In 2016, Moamen S. Refat *et al.* worked on Chromium(III) niacinamide compounds serving as a potential therapeutic model for diabetes mellitus. A novel binuclear Chromium(III)-niacinamide complex was synthesized by reacting a methanolic solution of chromium chloride and niacinamide at 60°C, yielding a compound exhibiting the empirical formula, $[Cr_2(Nic)(Cl)_6(H_2O)_4].H_2O$. The efficacy of the Chromium(III)-niacinamide compound in reducing blood glucose concentration and HbA1c was evaluated in diabetes-induced rats. Treatment with Nic sustained near-normal fasting blood glucose in STZ-induced diabetic animals. Remarkably, merging Nic with Cr^{3+} impressively heightened its glucose-lowering capacity, contrary to Nic alone. Since trivalent chromium plays a vital role in glucose metabolism, its fusion with Nic efficiently lessens blood glucose and HbA1C levels in diabetic rats. The Nic/C^{3+} compound also ameliorated gluconeogenic enzymes, refurbished liver glycogen to near-normal levels, and fortified antioxidant defense capacities (Fig. **15**) [28].

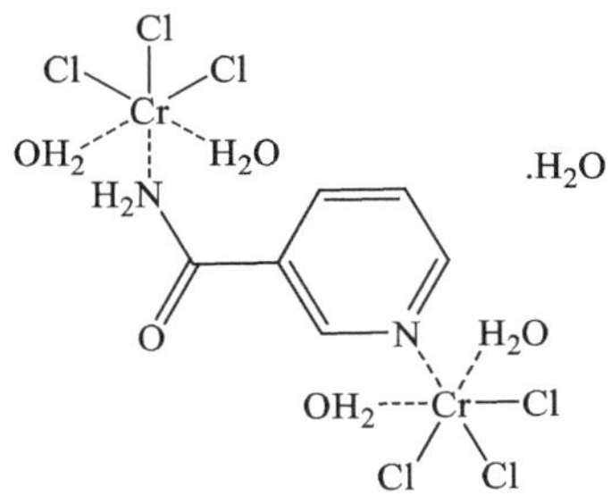

Fig. (15). Structure of $[Cr_2(Nic)(Cl)_6(H_2O)_4].H_2O$ complex.

In 2018, M. Lipko *et al.* conducted research to analyze the involvement of Reactive Oxygen Species (ROS) in glucose transport stimulated by chromium and insulin, using antioxidants to investigate the regulatory mechanisms. The effect of chromium ions on phosphatase enzymes associated with impaired insulin signaling has also been examined. In this research, it was established that ROS are crucial for the optimal metabolic effects of insulin and chromium(III) ions. Cr^{3+} ions amplified ROS levels in C2C12 mouse myotubes and boosted glucose transport, signifying a role of ROS in the metabolic action of this trace element. This was further supported by the use of antioxidants and DPI, which reduced the insulin/Cr3+-induced increase in glucose cross-membrane transport [29].

In 2019, Wei-Ling Guo *et al.* reported a cutting-edge Chromium(III) complex based on *Grifola frondosa* polysaccharide [GFP-Cr(III)]. The impact of [GFP-Cr(III)] complex on hyperglycemic and hyperlipidemic conditions was investigated in mice with diabetes induced by a high-fat diet and streptozotocin (STZ). These findings indicate that the complex may serve as a potential nutraceutical agent for managing elevated glucose and lipid levels. Cr(III), GFP, and the GFP-Cr(III) complex treatments crucially decreased fasting glucose levels compared to the DM group. At the same dose (300mg/kg/day), the GFP-Cr(III) complex displayed superior glucose-lowering effects compared to GFP alone in HFD and STZ-induced diabetic mice. In particular, high-dose supplementation (900 mg/kg/day) of GFP-Cr(III) complex induced the most striking effect, restoring fasting glucose levels and glucose tolerance to near-control levels. These findings indicate that Cr(III), particularly in the GFP-Cr(III) complex, enhances islet cell function and evidently improves glucose regulation in diabetic mice [30].

In 2020, Pengshou Li *et al.* successfully synthesized a complex of Chromium(III) derived from genistein (GEN), which exhibited considerable anti-hyperglycemic activity *in vivo* while showing no signs of toxicity. The proposed structure of this complex [$CrGEN_3$] included a chromium ion coordinated by three ligands. In the analysis of hyperglycemic activity, the complex enhanced glucose and lipid metabolism in db/db mice, which showed clinical features analogous to human type 2 diabetes, including polydipsia, polyphagia, polyuria, obesity, hyperglycemia, hyperinsulinemia, insulin resistance, and dyslipidemia. Treatment with complex notably reduced body weight ($p<0.05$), fasting and random blood glucose levels ($p<0.05$), and enhanced glucose and insulin tolerance in OGTT and ITT after 4 weeks ($p<0.05$). These outcomes indicate that the complex meritoriously lowers blood glucose and attenuates insulin resistance in diabetic mice (Fig. **16**) [31].

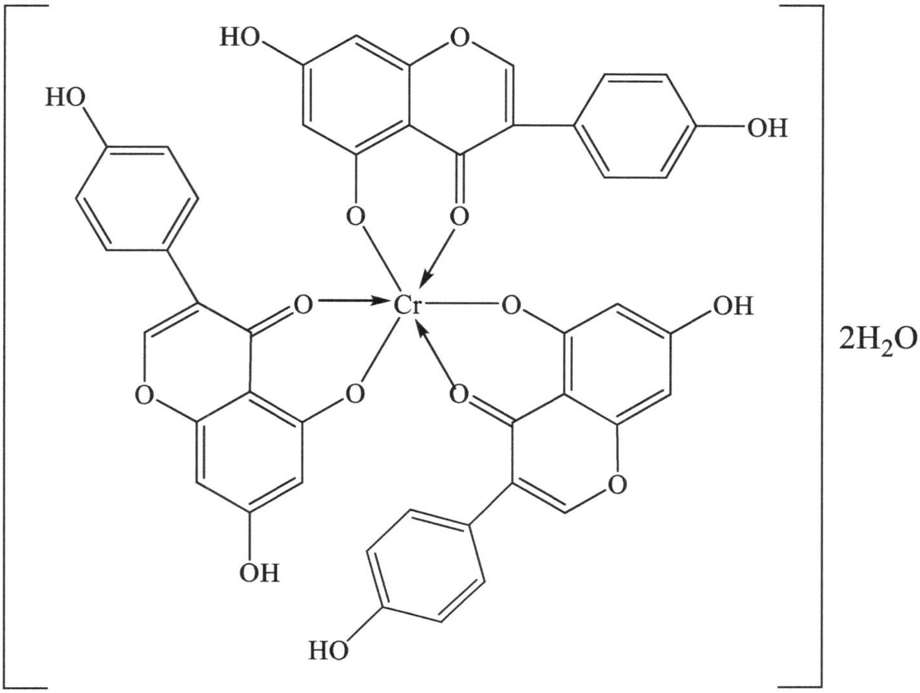

Fig. (16). The structure of genistein-Cr(III) complex.

In 2021, Jin-long Dong *et al.* synthesized a complex of Chromium(III), $Cr(C_7H_3NO_4)_2.C_4H_{12}N_5$by methanol-assisted chelation of chromium with dipicolinic (H_2dipic). The complex evinced satisfactory physicochemical characteristics and positive regulatory impact on glycemic levels and lipid profiles in type 2 diabetes mellitus patients. The dipicolinic-methanol-based complex showed *in vivo* efficacy against hyperglycemia and hyperlipidemia, with no signs of toxicity. Prefatory outcomes display that this complex lowered TC and TG levels by 35.38% and 32.40%, respectively, outperforming $CrCl^{3-}$ and Met-treated groups, probably because of the synergistic effects of Cr(III) with H_2dipic and Met. Acute oral toxicity tests revealed no lesions in the liver, kidney, or pancreas, although MTT assays indicated low cytotoxicity and decent biocompatibility. Thus, the Cr(III) complex stemming from H_2dipic and Met is non-toxic with potential as a safe hypoglycemic and lipid-lowering drug (Fig. **17**) [32].

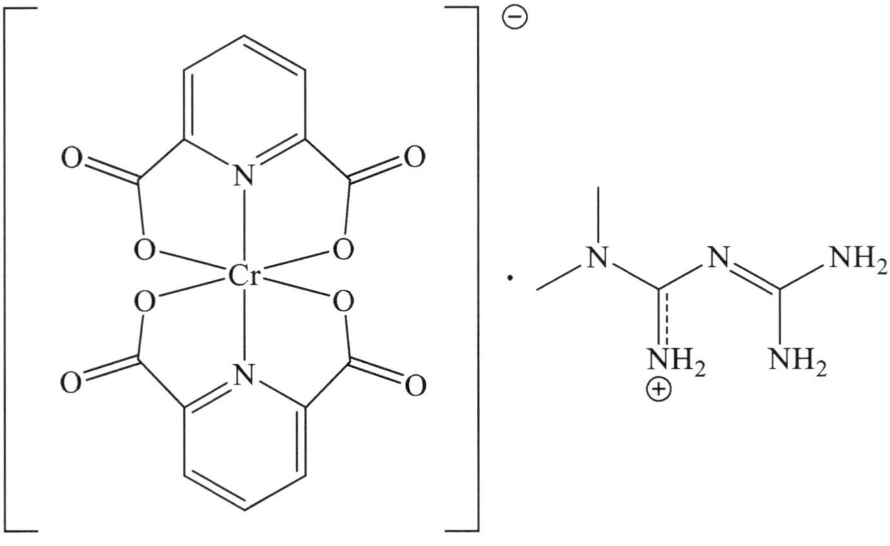

Fig. (17). Structure of [(HMet)[Cr(dipic)2].

In 2022, Mei-Ting Wang *et al.* investigated the inhibitory influence of Chromium(III)-fortified yeast (YCr) on hyperglycemia and hyperlipidemia triggered by a High-Fat, High-Fructose Diet (HFHFD) in mice, and moreover clarified its mechanistic insights considering the role of intestinal microbiota and hepatic metabolism.In this article, the precautionary effects of YCr and Cr intrusions on hyperlipidemia and hyperglycemia were inspected *via* intestinal microbiomics and liver metabolomics. Both intrusions considerably enhanced hyperglycemia, hyperlipidemia, and intestinal microbial disparity in HFHFD-fed mice, with YCr presenting superior effects. Liver metabolomics and RT-qPCR analyses further unveiled that YCr controlled metabolic profiles and the expression of key genes related to glucose and lipid metabolism. These conclusions underscore the potential of YCr as an efficient food for preventing hyperlipidemia and hyperglycemia [33].

In 2023, Xiaodong Ge *et al.*investigated the glucose-lowering underlying process of 6,8-guanidyl luteolin quinone-chromium co-ordination (GLQ.Cr), as well as assessed its impact on gastrointestinal microbiota in mice with T2DM. The GLQ.Cr attenuated signs of diabetes *via* improved hepatic and pancreatic activity, as well as tuning the gut microbiome composition (*Parabacteroides*, *Lactobacillus*, *Alistipes*, *Lachnoclostridium,* and *Desulfovibrio*). GLQ.Cr treatment ameliorated hyperglycemia, liver injury, tissue inflammation, and other T2DM symptoms. RNA-seq displayed alteration of abnormal glucose metabolism by means of PPAR signaling, with key genes activated at transcriptional and protein levels. Caecal content sequencing designated that GLQ.Cr dropped blood glucose by modulating intestinal microbes, a mechanism induced by fecal microbiota transplantation from treated mice, which enhanced microbial balance and lowered hyperglycemia. These verdicts accentuate gut microbiota regulation as a crucial anti-hyperglycemic mechanism of GLQ.Cr and suggest fecal transplantation as a potential tactic for studying flavonoid-based therapies in T2DM (Fig. **18**) [34].

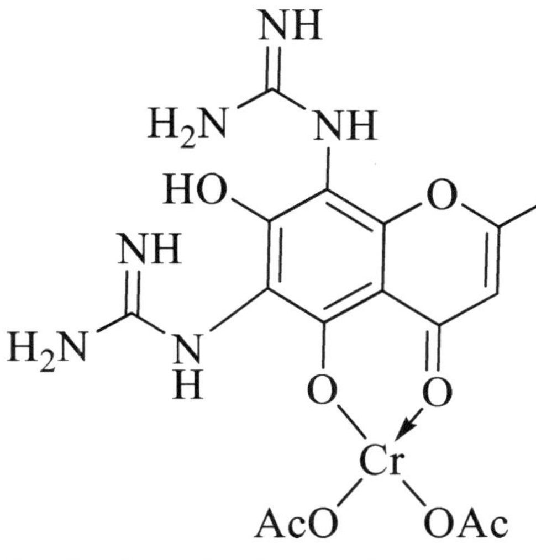

Fig. (18). 6,8-guanidyl luteolinquinone-chromium complex.

d. **Copper:** Copper is an essential cofactor in numerous metalloenzymes involved in cell metabolism. Copper has two oxidation states, +1 and +2, which permit it to catalyze redox reactions and produce Reactive Oxygen Species (ROS). Copper(I) ties up to thioether and thiol groups, while copper(II) associates with oxygen or imidazole nitrogen groups, influencing protein structures and biochemical processes. Even though toxic at high concentrations, copper and its complexes are much less hazardous than non-endogenous metals such as platinum, owing to their natural presence in the body.

In 2015, Sougata Ghosh *et al.* studied the biological method to prepare copper nanoparticles (CuNPs). *D. bulbifera* tuber extract (DBTE) was used to synthesize CuNPs that were additionally examined for anti-diabetic activity and reactive oxygen species (ROS) neutralizing capability. These were examined for downregulation of α-amylase and α-glucosidase. They stated the sturdy inhibitory activity of CuNPs against α-amylase and α-glucosidase, vital targets in T2DM treatment, accompanied by superior antioxidant potential. This is the first report of such activities for CuNPs produced *via* DBTE. Their elevated radical-scavenging and glycosidase inhibition *in vitro* afford strong evidence for the antidiabetic potential of CuNPs, upholding their remedial use in T2DM management [35].

In 2016, S. S. Laxmi *et al.* studied two novel Copper(II) complexes derived from Schiff base, [CuL^1(tmen)] and [Cu$_2$L$^2_2$(tmen)]. The intended complexes were examined for *in-vivo* anti-diabetic activity. The complexes were found to significantly hinder the functionality of α-glucosidase and α-amylase. This study assessed the inhibitory effects of Cu(II) complexes [CuL^1(tmen)] and [Cu$_2$L$^2_2$(tmen)] on α-amylase and α-glucosidase, vital enzymes in postprandial glucose regulation. [Cu$_2$L$^2_2$(tmen)] displayed average inhibition (α-amylase: 50.18±0.39%, α-glucosidase: 54.2±0.05%), while [CuL^1(tmen)] was feebler. Contrary to acarbose (~85%), [Cu$_2$L$^2_2$(tmen)] also revealed glucose diffusion more effectively than [CuL^1(tmen)], signifying delayed absorption. Theheightened activity of [Cu$_2$L$^2_2$(tmen)] may be ascribed to chloro substituents in its Schiff base ligand (Fig. **19**) [36].

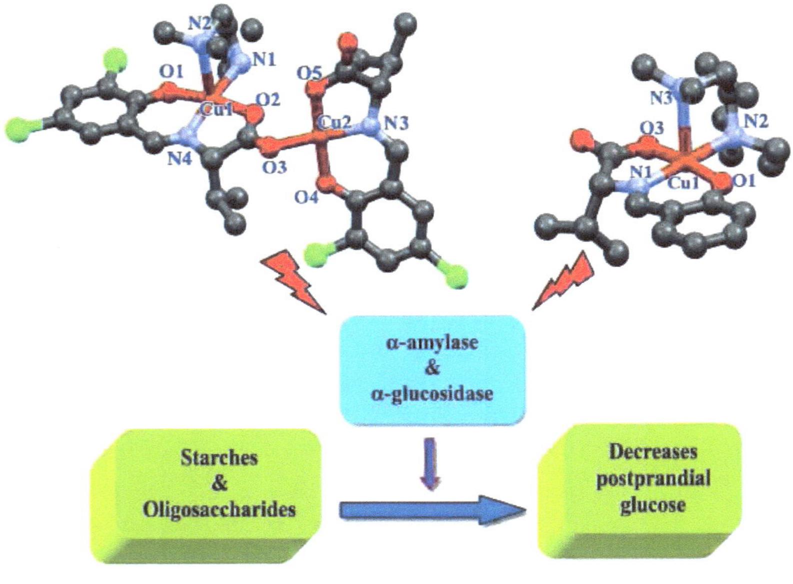

Fig. (19). [CuL^1(tmen)] and [Cu$_2$L$^2_2$(tmen)]{where, H$_2$L^1 = N-(salicylidene)-L-valine, H$_2$L^2 = N-(3,--dichlorosalicylidene)-L-valine and tmen = N,N,N,N-tetramethylethylene-1,2-diamine}.

In 2017, Rabia Javed *et al.* synthesized CuO NPs using the co-precipitation method and employed polyethylene-glycol (PEG) and polyvinyl-pyrrolidone (PVP) as surface-binding agents through physical adsorption.CuO-PEG and CuO-PVP nanoparticles revealed higher α-amylase inhibition (32.23% and 37.75%), contrary to uncapped CuO nanoparticles (19.17%). The heightened antidiabetic activity of capped nanoparticles is ascribed to reduced size and larger surface atom exposure, expanding surface area (Fig. **20**) [37].

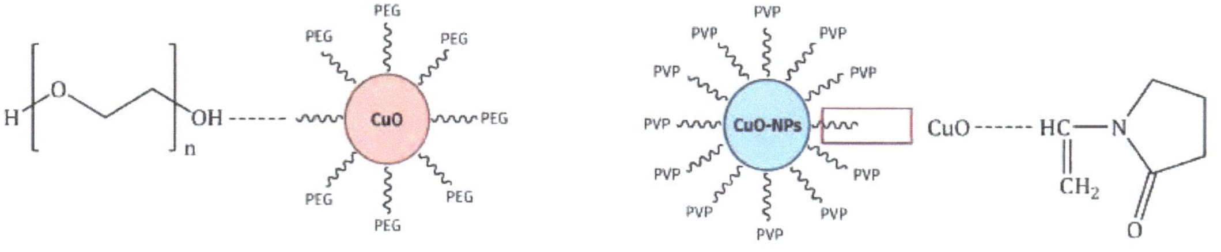

Fig. (20). Illustration of CuO nanoparticles stabilized with PEG and PVP.

In 2018, Tanay R. Bhagwat *et al.* outlined a quick and green synthesis route for copper nanoparticles utilizing three medicinal plants: leaf extract of *B. prionitis* (BPLE), *P. orientalis* (POLE), and peel extract of *L. chinensis* (LCPE),offering a sustainable process with dynamic biomedical applications. These CuNPs suppress the activity of both α-amylase and α-glucosidase enzymes, underscoring their potential as nanotherapeutic agents in clinical studies for type 2 diabetes mellitus. CuNPs fabricated by LCPE, BPLE, and POLE exhibited flexible enzyme inhibition. LCPE-CuNPs had the maximum α-amylase inhibition, trailed by BPLE and POLE, contrary to acarbose. For α-glucosidase, POLE-CuNPs revealed the utmost inhibition, followed by BPLE, although LCPE-CuNPs showed much lower activity [38].

In 2019, Afifa Mushtaq *et al.* studied four novel complexes of Copper(II): $[Cu_2(2\text{-}BrC_6H_4CH_2COO)_4(Phen)_2]$, $[Cu(2\text{-}BrC_6H_4CH_2COO)_2(Bipy)]$, $[Cu_2(2\text{-}BrC_6H_4CH_2COO)_4(3\text{-}BrPy)_2]$, and $[Cu_2(2\text{-}BrC_6H_4CH_2COO)_4(3\text{-}MePy)_2]$. The α-glucosidase and anticholinesterase inhibitory activities of the complexes were evaluated to explore their therapeutic potential in diabetes and Alzheimer's disease. The structure-activity correlation holds that the complexes [Cu2(2-BrC6H4CH2COO)4(Phen)2] and [Cu(2-BrC6H4CH2COO)2(Bipy)] hinder α-glucosidase by obstructing its active sites and inhibiting substrate access. Their nucleophilic and electrophilic centers enable reversible binding *via* weak non-covalent interactions. The metal center interacts with nucleophilic sites; meanwhile, carboxylate ligands form hydrogen bonds with electrophilic sites. On the other hand, the paddlewheel structures of $[Cu_2(2\text{-}BrC_6H_4CH_2COO)_4(3\text{-}BrPy)_2]$ and $[Cu_2(2\text{-}BrC_6H_4CH_2COO)_4(3\text{-}MePy)_2]$restricted such interactions, exhibiting no inhibition. The IC_{50} values of complexes $[Cu_2(2\text{-}BrC_6H_4CH_2COO)_4(Phen)_2]$ and $[Cu(2\text{-}BrC_6H_4CH_2COO)_2(Bipy)]$ were higher than formerly reported Cu(II) complexes (Fig. **21**) [39].

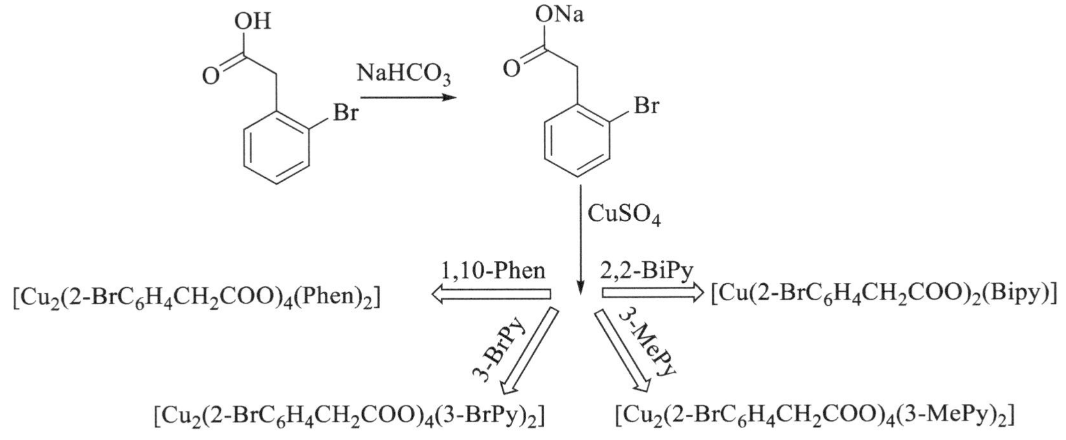

Fig. (21). Synthesis scheme for complexes.

In 2020, Sadaf Noor and colleagues conducted a study on a simplified mycogenic synthesis approach for CuNPs production. The preparation of CuNPs was implemented via the *Aspergillus niger* strain STA9, which had undergone both molecular and morphological characterization. Fungus-synthesized CuNPs suppressed α-glucosidase, confirming their antidiabetic potential. Cu^{2+} ions and their compounds have presented sturdy *in vitro* α-glucosidase inhibition, in certain cases surpassing that of acarbose, a standard antidiabetic drug [40].

In 2021, Shifa J. Ali and colleagues carried out the current research aimed at fabricating CuNPs derived from *Mucuna pruriens*in addition to examining their oxidative stress-reducing and blood glucose-regulating activity. The antidiabetic potential was assessed through an α-amylase inhibitory assay, measuring the percentage inhibition exhibited by CuNPs developed from Mucuna pruriens. The inhibition was 66% for 10 μL, 69% for 20 μL, 73% for 30 μL, 79% for 40 μL, and 80% for 50 μL. The outcomes designate that CuNPs inhibit α-amylase, affording a therapeutic advantage in diabetes by regulating blood glucose levels. Chronic hyperglycemia produces ROS, leading to oxidative stress, lipid peroxidation, and membrane impairment. Consequently, precluding oxidative damage with free radical scavengers and preventing digestive enzymes like α-amylase and α-glucosidase are vital approaches for diabetes management [41].

In 2022, Shaik Ameena *et al*. reported that CuNPs, prepared by reducing 3mL of aqueous copper acetate using an aqueous extract derived from *Cocculus hirsutus* foliage, exhibited promising antioxidant, antibacterial, and antidiabetic activities. This work suggests that CH-CuNPs can be fabricated as an alternative antidiabetic agent. Enzyme inhibition assays exhibited dose-dependent activity, with α-amylase suppression extending from 25.6% to 64.5%, and α-glucosidase suppression from 20.6% to 68.5%. Particularly, CH-CuNPs unveiled sturdier inhibition of α-glucosidase than α-amylase [42].

In 2023, Tunde L. Yusuf and colleagues synthesized three Cu(II) complexes from Schiff base ligands bearing halogen-substituted ON donor groups, derived from the reaction of 2-hydroxybenzaldehyde with three different halogenated anilines. These complexes were assessed for anti-diabetic (α-amylase as well as α-glucose enzymes inhibition assay) activity. Free ligands and their copper complexes CuL1 and CuL2 displayed weaker α-glucosidase inhibition than the control, while CuL3 was more effective atall concentrations, along with its heightened potency allied to the $-NO_2$ group. For α-amylase, the control was commonly sturdier, though free ligands outpaced their complexes. Peculiarly, L3 at 1000 μM surpassed the control; L1 and L2 were weaker at 250-500 μM but comparable at 1000 μM (Fig. **22**) [43].

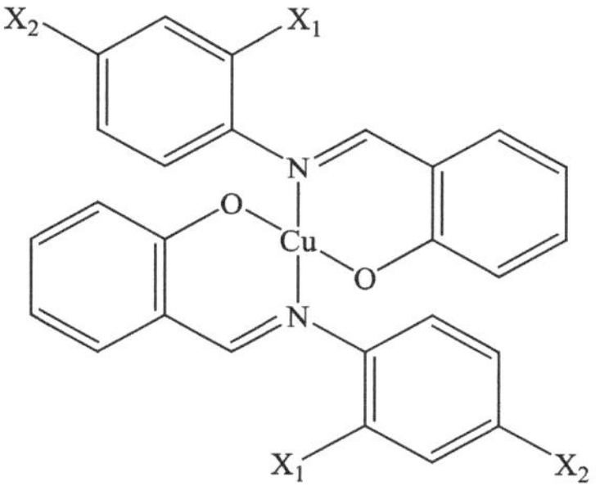

X_1 = Br, X_2 = CuL1; X_1 = Br, X_2 = CH_3:CuL2; and X_1 = I, X_2 = NO_2:CuL3

Fig. (22). Cu(II) complexes from Schiff base ligands bearing halogen-substituted ON donor groups.

In 2024, Ankush Relhan *et al.* prepared CuO nanoparticles (CuO-NPs) utilizing foliage extracts from Allahabad Safeda (1-9%) and Hisar Safeda (3-11%), with the goal of promoting health and environmental safety. The nanoparticles were then assessed for antidiabetic efficacy through α-glucosidase and α-amylase enzyme inhibitory bioassays. In α-amylase inhibition, acarbose displayed 77.98% inhibition. CuO-NPs from Allahabad Safeda unveiled 67.32% inhibition, remarkably greater than Hisar Safeda (63.25%). For α-glucosidase, acarbose reached 80.56% inhibition, whereas Allahabad Safeda CuO-NPs presented 75.18% outpacing Hisar Safeda (71.38%). Both extracts, therefore, displayed sturdy inhibitory potential (Fig. **23**) [44].

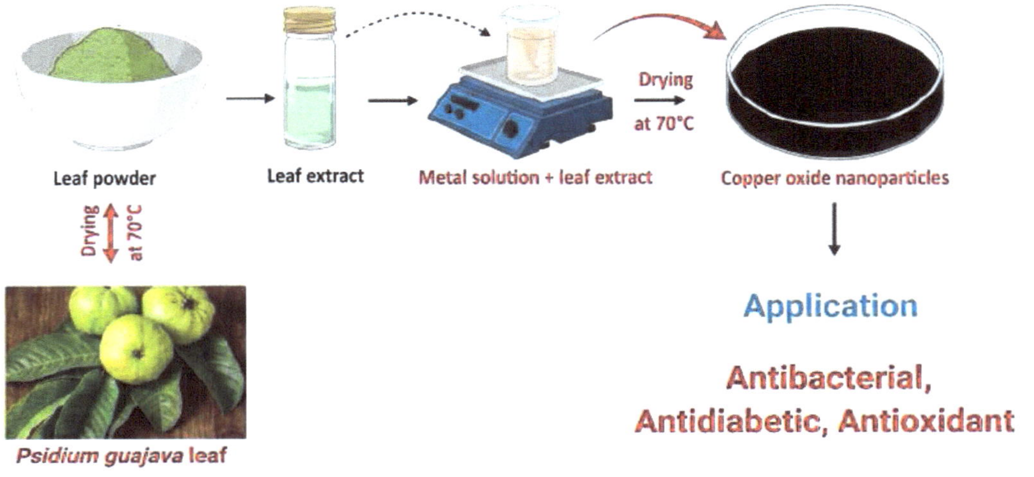

Fig. (23). CuO nanoparticles (CuO-NPs) utilizing foliage extracts from Allahabad Safeda and Hisar Safeda.

e. **Cobalt:** Cobalt is a chemically reactive and potentially hazardous metal that is not found naturally in its metallic state. It is infrequent but found in mineral water, seawater, and certain rocks. It is acquired from ores like cobaltite, erythrite, esmaltite, and linneite. Most stable oxidation states of cobalt are +2 and +3 [45]. Cobalt is endowed with various characteristics, and such a range makes it a promising candidate for drug development [46].

In 2015, A. H. Manikshete *et al.* synthesized a novel ligand, salicylaldehyde- 4-chlorobenzoylhydrazone (HL), and its metal complexes of Mn(II), Co(II), and Ni(II). The anti-hyperglycemic potential of the produced complexes was assessed *in vitro* through an assay for amylase inhibitory activity employing a DNSA-based colorimetric assay (Fig. **24**) [47].

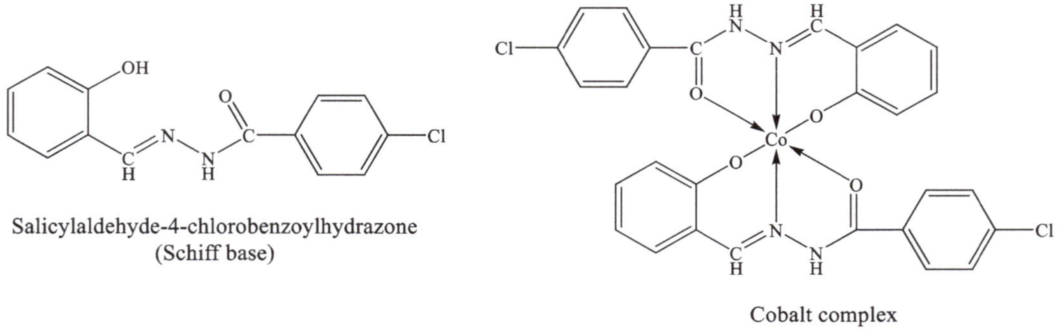

Fig. (24). Salicylaldehyde-4-chlorobenzoylhydrazone and its Cobalt complex.

In 2016, I. P. Tripathi and Aparna Dwivedi developed N-methylenediamine-based metal [Cu(II), Co(II), Ni(II), Zn(II)] complexes. In this work, they produced copper, cobalt, nickel, and zinc complexes with N-methylethylenediamine to gauge their α-glucosidase inhibitory activity. Entirely twelve complexes revealed inhibition, with $[Co(men)_3]2NO_3$ unveiling the sturdiest effect and $[Ni(men)_3]2Cl$ the feeblest (Fig. **25**) [48].

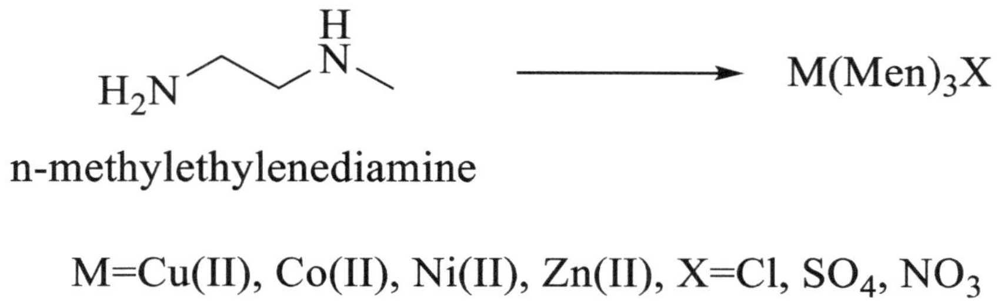

Fig. (25). N-methyethylenediamine-based metal [Cu(II), Co(II), Ni(II), Zn(II)] complexes.

In 2017, I. E. Otuokere and K. C. Amadi synthesized a cobalt complex of glimepiride by reacting glimepiride with Cobalt(II) chloride hexahydrate. The cobalt-glimepiride complex exhibited remarkably sturdier hypoglycemic activity than glimepiride single-handedly in alloxan-induced albino rats. 8 hours later, the complex lowered blood glucose from217.50±21.93 to 134.25±10.90 mg/dL, contrary to glimepiride (258.25±25.38 to 187.75±18.71 mg/dL). Even after 21 days, the complex upheld a better effect, dropping glucose from 121.25±1.50 to 93.00±3.46 mg/dL, whereas glimepiride reduced it from 122.50±2.51 to 100.25±2.87 mg/dL. These fallouts underscore the heightened and persistent hypoglycemic potential of the metal complex (Fig. **26**) [49].

In 2019, Xia Wang *et al.* synthesized a three-dimensional Co(II) complex $\{[Co_3(TTTMB)_4(SCN)_6]\cdot 3.5H_2O\}_n$ using a conformationally adaptable tripodal ligand KSCN, 1,3,5-tris(triazol-1-ylmethyl)-2,4,6-trimethylbenzene (TTTMB), and $Co(NO_3)_2$. The complex revealed sturdy, dose-dependent inhibition of α-amylase, presenting a clear linear relationship between concentration and activity. Inhibition amplified from 3.28% to 94.19% across 0.1-2.5 mg/mL, representing its potent effect. In contrast to the reference drug acarbose, the complex displayed a lower IC_{50} value, authorizing sturdier inhibitory activity. Fascinatingly, the free ligand TTTMB exhibited no inhibitory effect, but coordination with Co(II) notably enhanced activity, underlining the vital role of the metal center in the biological utility of the complex (Fig. **27**) [50].

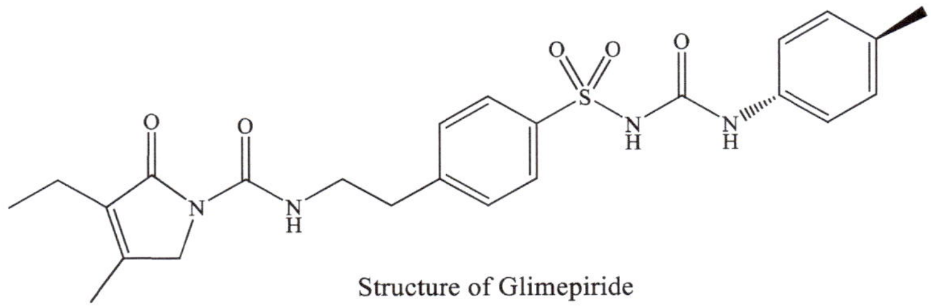

Structure of Glimepiride

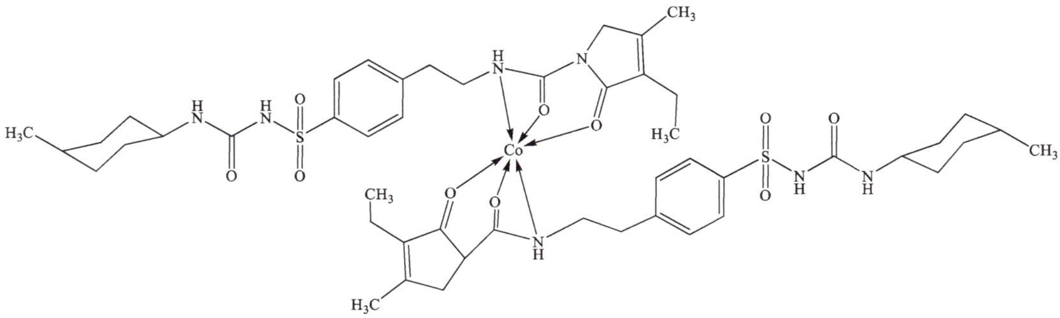

Cobalt(II) Glimepiride complex

Fig. (26). Glimepiride and its Cobalt(II) complex.

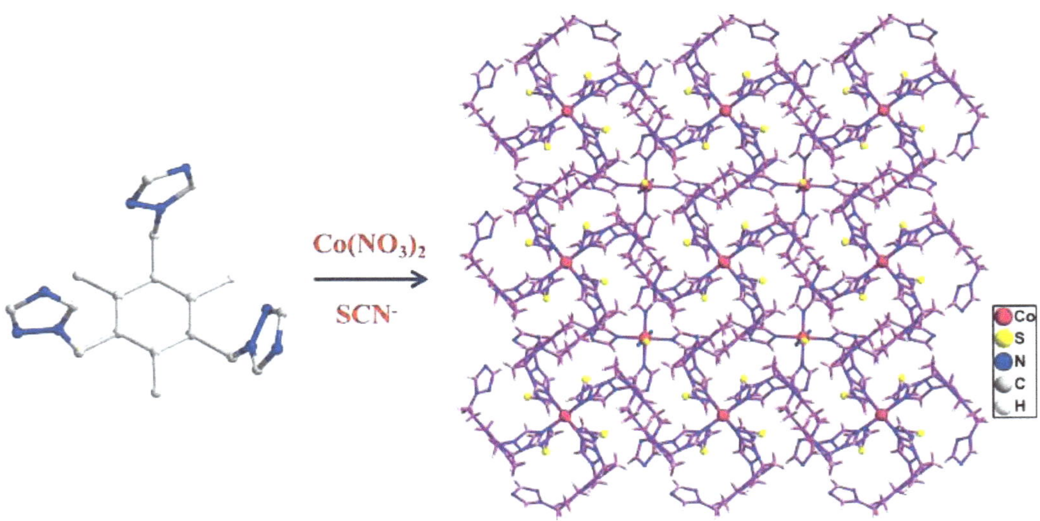

Fig. (27). Three-dimensional Co(II) complex $\{[Co_3(TTTMB)_4(SCN)_6]\cdot 3.5H_2O\}_n$.

In 2020, Md. M. Hassanein and colleagues investigated the antidiabetic activities of flavonoid-based Cobalt-Quercetin Complex (CQC) and compared them to those of insulin in diabetic rats, to assess its efficacy as a potential therapeutic agent. CQC effectively ameliorated diabetes-induced metabolic disturbances, primarily *via* its strong antioxidant potential. The hypoglycemic effect of CQC became apparent from day 7 of treatment and was more prominent by the final day, signifying that CQC may perform by suppressing intestinal glucose absorption and/or provocating insulin release, possibly through β-cell regeneration and functional enhancement. Biochemical analysis revealed that the hypoglycemic effects of CQC and insulin were almost analogous, with FBG declines of 51% and 53%, respectively. These outcomes aligned with histopathological clarifications; insulin-treated rats unveiled intact islets of Langerhans with mild β-cell degeneration, while CQC treatment exhibited slight islet atrophy but no β-cell degeneration. This provides evidence of CQC's pancreatic protective and curative effects, favoring its antidiabetic potential (Fig. **28**) [51].

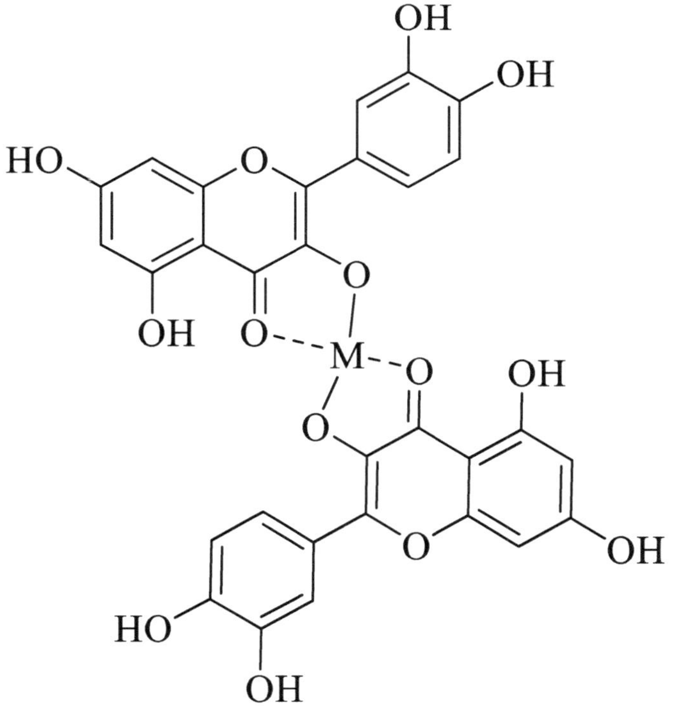

Fig. (28). Flavonoid-based cobalt-quercetin complex.

In 2021, R. V. Sakhtivel *et al.* synthesized novel metal (Co, Ni, Cu) complexes derived from Schiff base ligand of 2,2'-((1E,1'E)-((4-nitro-

1,2-phenylene)bis(azanylylidene))bis(methanylylidene))diphenol. An α-amylase inhibition assay was used to evaluate antidiabetic activity. The inhibition of α-amylase, a vital therapeutic approach for lowering postprandial blood glucose, was gauged for the synthesized complexes at diverse concentrations (50, 100, 200, 250, 500 mM). Fallouts discovered that the copper complex unveiled greater antidiabetic activity, contrary to the nickeland cobalt complexes, with deviations ascribed to the intrinsic properties of the metal ions (Fig. **29**) [52].

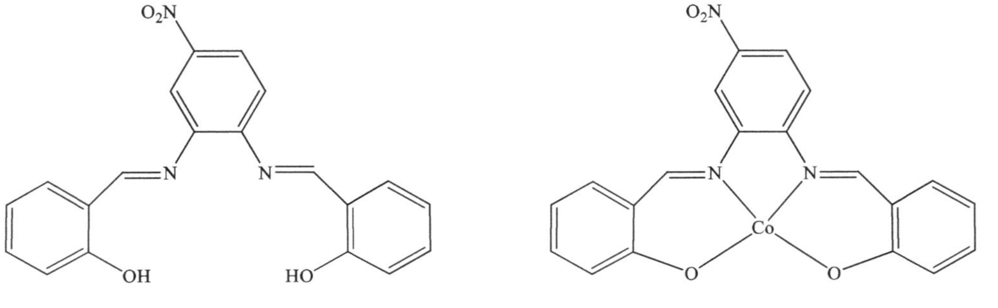

2,2'-((1E,1'E)-((4-nitro-1,2-phenylene)bis(azanylylidene))bis (methanylylidene)) diphenol

Cobalt complex

Fig. (29). 2,2'-((1E,1'E)-((4-nitro-1,2-phenylene)bis(azanylylidene))bis(methanylylidene))diphenol and its Cobalt complex.

In 2023, Aliya Ajaz *et al.* synthesized complexes of Zn(II), Co(II), Cr(III), and VO(IV) using a Schiff base ligand, (E)-2-((adamantan-1-ylimino)methyl)- 6-allylphenol, prepared by condensing amantadine with 3-allyl-2-hydroxybenzaldehyde under reflux conditions. Evaluation of these metal complexes was carried out to determine their therapeutic activity against diabetes *via* α-glucosidase and α-amylase *in vitro* inhibition studies. α-glucosidase inhibition was assessed at 30.41, 60.56, and 90.90 μg/mL, although α-amylase inhibition was measured at 60.22, 120.71, and 171.45 μg/mL, presenting a sturdy dose-dependent response. Out of the complexes, Co(II) exhibited the utmost α-glucosidase inhibition, trailed by Cr(III), VO(IV), and Zn(II). On the contrary, VO(IV) unveiled the resilient α-amylase inhibition, trailed by Zn(II), Cr(III), and Co(II) (Fig. **30**) [53].

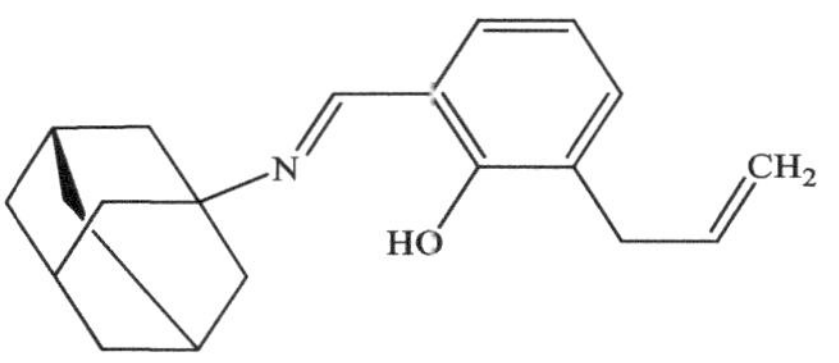

(E)-2-((adamantan-1-ylimino)methyl)-6-allylphenol

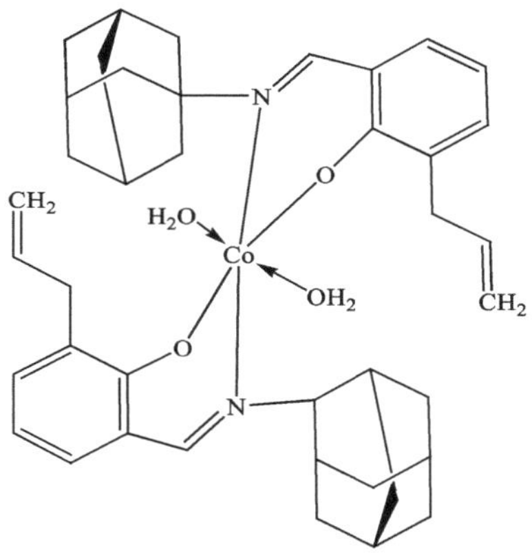

Cobalt complex

Fig. (30). (E)-2-((adamantan-1-ylimino)methyl)-6-allylphenol and its cobalt complex.

In 2024, Shivakumar Venkataramaiah *et al.* reported the phytogenic synthesis of cobalt ferrite nanoparticles (DHLE.$CoFe_2O_4$NPs) using *Decalepis hamiltonii* leaf extract. These nanoparticles demonstrated potent inhibition against α-glucosidase and α-amylase, crucial targets for managing diabetes, in a dose-dependent manner. At 200 μg/mL, the NPs reached 46.45% α-amylase inhibition, compared with 65.52% for acarbose. Correspondingly, they exhibited 48.66% α-glucosidase inhibition, adjacent to acarbose (65.05%). Additionally, they were effective against oxidative stress-induced inflammation, diabetes, and thrombosis [54].

CONCLUSION

In the given chapter, we have discussed the antidiabetic medications made up of metal complexes. The increasing prevalence of diabetes worldwide underlines the importance of developing novel medications, involving metal-based therapies. Metal complexes, significantly those including vanadium and zinc, have been contemplated as mimics of insulin activity, which could substitute for injections. These compounds work by modulating blood glucose *via* the interaction with insulin-signaling pathways, consequently being therapeutic agents. The evolution of such complexes paves the way to tackle diabetes more effectively, merging biochemical insights with innovative approaches in medicinal chemistry.

REFERENCES

[1] Global Burden of Disease Collaborative Network. (2024), Retrieved from Global Burden of Disease Study 2021. Results. Institute for Health Metrics and Evaluation.

[2] Hossain MJ, Al-Mamun M, Islam MR. Diabetes mellitus, the fastest growing global public health concern: Early detection should be focused. Health Sci Rep 2024; 7(3): e2004. [http://dx.doi.org/10.1002/hsr2.2004] [PMID: 38524769]

[3] Ong KL, Stafford LK, McLaughlin SA, *et al.* Global, regional, and national burden of diabetes from 1990 to 2021, with projections of prevalence to 2050: a systematic analysis for the Global Burden of Disease Study 2021. Lancet 2023; 402(10397): 203-34. [http://dx.doi.org/10.1016/S0140-6736(23)01301-6] [PMID: 37356446]

[4] Ahmad E, Lim S, Lamptey R, Webb DR, Davies MJ. Type 2 diabetes. Lancet 2022; 400(10365): 1803-20. [http://dx.doi.org/10.1016/S0140-6736(22)01655-5] [PMID: 36332637]

[5] Lovic D, Piperidou A, Zografou I, Grassos H, Pittaras A, Manolis A. The growing epidemic of diabetes mellitus. Curr Vasc Pharmacol 2020; 18(2): 104-9. [http://dx.doi.org/10.2174/1570161117666190405165911] [PMID: 30961501]

[6] Rajput R, YogeshYadav , Rajput M, Nanda S. Utility of HbA1c for diagnosis of gestational diabetes mellitus. Diabetes Res Clin Pract 2012; 98(1): 104-7. [http://dx.doi.org/10.1016/j.diabres.2012.02.018] [PMID: 22456454]

[7] Khunti K, Valabhji J, Misra S. Diabetes and the COVID-19 pandemic. Diabetologia 2023; 66(2): 255-66. [http://dx.doi.org/10.1007/s00125-022-05833-z] [PMID: 36418578]

[8] Sarker A, Das R, Ether S, Saif-Ur-Rahman KM. Non-pharmacological interventions for the prevention of type 2 diabetes mellitus in low and middle-income countries: protocol for a systematic review and meta-analysis of randomized controlled trials. Syst Rev 2020; 9(1): 288. [http://dx.doi.org/10.1186/s13643-020-01550-z] [PMID: 33298172]

[9] Pan XR, Li GW, Hu YH, *et al.* Effects of diet and exercise in preventing NIDDM in people with impaired glucose tolerance. The Da Qing IGT and Diabetes Study. Diabetes Care 1997; 20(4): 537-44. [http://dx.doi.org/10.2337/diacare.20.4.537] [PMID: 9096977]

[10] Harris MI, Eastman RC. Early detection of undiagnosed diabetes mellitus: a US perspective. Diabetes Metab Res Rev 2000; 16(4): 230-6. [http://dx.doi.org/10.1002/1520-7560(2000)9999:9999<::AID-DMRR122>3.0.CO;2-W] [PMID: 10934451]

[11] Lawrance; Geoffrey A. Introduction to Coordination Chemistry. Wiley, 2010. [http://dx.doi.org/10.1002/9780470687123]

[12] Greenwood NN, Earnshaw A. Chemistry of the Elements. 2nd ed., Butterworth-Heinemann 1997.bAmaral LMPF, Moniz T, Silva AMN, Rangel M. Vanadium Compounds with Antidiabetic Potential. Int J Mol Sci 2023; 24(21): 15675. [http://dx.doi.org/10.3390/ijms242115675] [PMID: 37958659]

[13] Rafique S, Idrees M, Nasim A, Akbar H, Athar A. Transition metal complexes as potential therapeutic agents. Biotechnol Mol Biol Rev 2010; 5(2): 38-45.

[14] Crans DC. Antidiabetic, chemical, and physical properties of organic vanadates as presumed transition-state inhibitors for phosphatases. J Org Chem 2015; 80(24): 11899-915. [http://dx.doi.org/10.1021/acs.joc.5b02229] [PMID: 26544762]

[15] Phanse MA, Patil MJ, Abbulu K. Synthesis, characterization and evaluation of the suppression of insulin resistance in Type-II diabetes mellitus animals by treatment with metal complex. Saudi J Biol Sci 2016; 23(3): 420-5. [http://dx.doi.org/10.1016/j.sjbs.2015.08.017] [PMID: 27081369]

[16] Jakusch T, Kiss T. *In vitro* study of the antidiabetic behavior of vanadium compounds. Coord Chem Rev 2017; 351: 118-26. [http://dx.doi.org/10.1016/j.ccr.2017.04.007]

[17] Patel N, Prajapati AK, Jadeja RN, *et al.* Model investigations for vanadium-protein interactions: Synthesis, characterization and antidiabetic properties. Inorg Chim Acta 2019; 493: 20-8. [http://dx.doi.org/10.1016/j.ica.2019.04.050]

[18] Patel N, Prajapati AK, Jadeja RN, *et al.* Dioxidovanadium(V) complexes of a tridentate ONO Schiff base ligand: Structural characterization, quantum chemical calculations and *in-vitro* antidiabetic activity. Polyhedron 2020; 180: 114434. [http://dx.doi.org/10.1016/j.poly.2020.114434]

[19] Semiz S. Vanadium as potential therapeutic agent for COVID-19: A focus on its antiviral, antiinflamatory, and antihyperglycemic effects. J Trace Elem Med Biol 2022; 69: 126887. [http://dx.doi.org/10.1016/j.jtemb.2021.126887] [PMID: 34798510]

[20] Xia J, Yu J, Xu H, *et al.* Comparative effects of vitamin and mineral supplements in the management of type 2 diabetes in primary care: A systematic review and network meta-analysis of randomized controlled trials. Pharmacol Res 2023; 188: 106647. [http://dx.doi.org/10.1016/j.phrs.2023.106647] [PMID: 36638933]

[21] Jurowska A, Hodorowicz M, Arabasz M, Szklarzewicz J. Protonated amino alcohols as cations in synthesis of vanadium(V) complexes with Schiff base ligands as potential insulin-mimetic agents. J Mol Struct 2024; 1305: 137722. [http://dx.doi.org/10.1016/j.molstruc.2024.137722]

[22] Holleman A F, Wiberg E, Wiberg N. "Zink", Lehrbuch der Anorganischen Chemie (in German) (91–100 ed.), 1985, Walter de Gruyter.

[23] Saito R, Tamura M, Kawano S, *et al.* Synthesis and biological evaluation of 4-hydroxy-5-oxo-2,5-dihydro-1H-pyrrole-3-carboxamides and their zinc (II) complexes as candidate antidiabetic agents. New J Chem 2017; 41(13): 5572-81. [http://dx.doi.org/10.1039/C7NJ00970D]

[24] Koles T. Dobravc, Maejima K, Yoshikawa Y, Meden A, Yasuib H, Perdih F. Bis(picolinato) complexes of vanadium and zinc as potential antidiabetic agents: synthesis, structural potential antidiabetic agents: synthesis, structural study. New J Chem 2018; 3619. [http://dx.doi.org/10.1039/C7NJ04189F]

[25] Govindan N, Vairaprakasam K, Chinnasamy C, Sivalingam T, Mohammed MKA. Green synthesis of Zn-doped *Catharanthus Roseus* nanoparticles for enhanced anti-diabetic activity. Mater Adv 2020; 1(9): 3460-5. [http://dx.doi.org/10.1039/D0MA00698J]

[26] Esteban-Parra GM, Sebastián ES, Cepeda J, *et al.* Anti-diabetic and anti-parasitic properties of a family of luminescent zinc coordination compounds based on the 7-amino-5-methyl-1,2,4-triazol-[1,5-a]pyrimidine ligand. J Inorg Biochem 2020; 212: 111235. [http://dx.doi.org/10.1016/j.jinorgbio.2020.111235] [PMID: 32920434]

[27] Mahmoud MA, Zaitone SA, Ammar AM, Sallam SA. Synthesis, structure and antidiabetic activity of chromium(III) complexes of metformin Schiff-bases. J Mol Struct 2015. [http://dx.doi.org/10.1016/j.molstruc.2015.11.055]

[28] Refat MS, El-Megharbel SM, Hussien MA, *et al.* Spectroscopic, structural characterizations and antioxidant capacity of the chromium (III) niacinamide compound as a diabetes mellitus drug model. Spectrochim Acta A Mol Biomol Spectrosc 2017; 173: 122-31. [http://dx.doi.org/10.1016/j.saa.2016.08.053] [PMID: 27619974]

[29] Lipko M, Debski B. Mechanism of insulin-like effect of chromium(III) ions on glucose uptake in C2C12 mouse myotubes involves ROS formation. J Trace Elem Med Biol 2018; 45: 171-5. [http://dx.doi.org/10.1016/j.jtemb.2017.10.012] [PMID: 29173475]

[30] Guo WL, Shi FF, Li L, *et al.* Preparation of a novel Grifola frondosa polysaccharide-chromium (III) complex and its hypoglycemic and hypolipidemic activities in high fat diet and streptozotocin-induced diabetic mice. Int J Biol Macromol 2019; 131: 81-8. [http://dx.doi.org/10.1016/j.ijbiomac.2019.03.042] [PMID: 30851330]

[31] Li P, Cao Y, Song G, *et al.* Anti-diabetic properties of genistein-chromium (III) complex in db/db

diabetic mice and its sub-acute toxicity evaluation in normal mice. J Trace Elem Med Biol 2020; 62: 126606.
[http://dx.doi.org/10.1016/j.jtemb.2020.126606] [PMID: 32650064]

[32] Dong J, Wen B, Song Z, *et al.* Potential antidiabetic molecule involving a new chromium(III) complex of dipicolinic and metformin as a counter ion: Synthesis, structure, spectroscopy, and bioactivity in mice. Arab J Chem 2021; 14(7): 103236.
[http://dx.doi.org/10.1016/j.arabjc.2021.103236]

[33] Wang MT, Guo WL, Yang ZY, *et al.* Intestinal microbiomics and liver metabolomics insights into the preventive effects of chromium (III)-enriched yeast on hyperlipidemia and hyperglycemia induced by high-fat and high-fructose diet. Curr Res Food Sci 2022; 5: 1365-78.
[http://dx.doi.org/10.1016/j.crfs.2022.08.015] [PMID: 36092021]

[34] Ge X, He X, Liu J, *et al.* Amelioration of type 2 diabetes by the novel 6, 8-guanidyl luteolin quinone-chromium coordination *via* biochemical mechanisms and gut microbiota interaction. J Adv Res 2023; 46: 173-88.
[http://dx.doi.org/10.1016/j.jare.2022.06.003] [PMID: 35700921]

[35] Ghosh S, More P, Nitnavare R, Jagtap S, Chippalkatti R, Derle A, Kitture R, Asok A, Kale S, Singh S, Shaikh M. L, Ramanamurthy B, Bellare J, Chopade B. A. Antidiabetic and antioxidant properties of copper nanoparticles synthesized by medicinal plant dioscorea bulbifera. Journal of Nanomedicine & Nanotechnology S6, 2015; pp. 007.
[http://dx.doi.org/10.4172/2157-7439.S6-007]

[36] Lakshmi SS, Geetha K, Gayathri M, Shanmugam G. Synthesis, crystal structures, spectroscopic characterization and *in vitro* antidiabetic studies of new Schiff base Copper(II) complexes. J Chem Sci 2016; 128(7): 1095-102.
[http://dx.doi.org/10.1007/s12039-016-1099-8]

[37] Javed R, Ahmed M, Haq I, Nisa S, Zia M. PVP and PEG doped CuO nanoparticles are more biologically active: Antibacterial, antioxidant, antidiabetic and cytotoxic perspective. Mater Sci Eng C 2017; 79(79): 108-15.
[http://dx.doi.org/10.1016/j.msec.2017.05.006] [PMID: 28628996]

[38] Bhagwat TR, Joshi KA, Parihar VS, Asok A, Bellare J, Ghosh S. Biogenic Copper Nanoparticles from Medicinal Plants as novel anti-diabetic nanomedicines. World J Pharm Res 2018; 7: 183-96.

[39] Afifa Mushtaq , Ali S, Tahir MN, Haider A, Ismail H, Iqbal M. Mixed-ligand Cu(II) carboxylates: Synthesis, crystal structure, FTIR, DNA binding, antidiabetic, and anti-alzheimer's studies. Russ J Inorg Chem 2019; 64(11): 1365-78.
[http://dx.doi.org/10.1134/S0036023619110147]

[40] Noor S, Shah Z, Javed A, *et al.* A fungal based synthesis method for copper nanoparticles with the determination of anticancer, antidiabetic and antibacterial activities. J Microbiol Methods 2020; 174: 105966.
[http://dx.doi.org/10.1016/j.mimet.2020.105966] [PMID: 32474053]

[41] Ali SJ, Geetha RV, Rajeshkumar S. Biosynthesis of copper nanoparticles using mucuna pruriens and its antioxidant and antidiabetic activity. J Pharm Res Int 2021; 33(47B): 621-9.
[http://dx.doi.org/10.9734/jpri/2021/v33i47B33163]

[42] Ameena S, Rajesh N, Anjum SM, *et al.* Antioxidant, antibacterial, and anti-diabetic activity of green synthesized copper nanoparticles of cocculus hirsutus (Menispermaceae). Appl Biochem Biotechnol 2022; 194(10): 4424-38.
[http://dx.doi.org/10.1007/s12010-022-03899-4] [PMID: 35357664]

[43] Yusuf TL, Waziri I, Olofinsan KA, Akintemi EO, Hosten EC, Muller AJ. Evaluating the *in vitro* antidiabetic, antibacterial and antioxidant properties of copper(II) Schiff base complexes: An experimental and computational studies. J Mol Liq 2023; 389: 122845.
[http://dx.doi.org/10.1016/j.molliq.2023.122845]

[44] Relhan A, Guleria S, Bhasin A, Mirza A, Zhou JL. Biosynthesized copper oxide nanoparticles by Psidium guajava plants with antibacterial, antidiabetic, antioxidant, and photocatalytic capacity. Biomass Conversion and Biorefnery 2024.
[http://dx.doi.org/10.1007/s13399-024-05544-y]

[45] Sotiles AR, Massarotti F, Pires JCO, Facchin Ciceri ME, Budziak Parabocz CR. Cobalt complexes: Introduction and spectra analysis. Orbital: The Electronic Journal of Chemistry 2019; 11(6).
[http://dx.doi.org/10.17807/orbital.v11i6.1242]

[46] Heffern MC, Yamamoto N, Holbrook RJ, Eckermann AL, Meade TJ. Cobalt derivatives as promising therapeutic agents. Curr Opin Chem Biol 2013; 17(2): 189-96.
[http://dx.doi.org/10.1016/j.cbpa.2012.11.019] [PMID: 23270779]

[47] Manikshete AH, Awatade MM, Sarsamkar SK, Asabe MR. Synthesis, characterization, antimicrobial, anticancer and antidiabetic activity of new Manganese (II), Nickel (II) and Cobalt (II) complexes with Salicylaldehyde-4-chlorobenzoylhydrazone. Int J Eng Sci Invent 2015; 4(1): 22-9.

[48] Tripathi I, Dwivedi A. Synthesis, characterization and α- glucosidase inhibition of some copper, cobalt, nickel and zinc complexes with n – methylethylenediamine. Br J Med Med Res 2016; 16(6): 1-11.
[http://dx.doi.org/10.9734/BJMMR/2016/26100]

[49] Otuokere IE, Amadi KC. Synthesis, characterization and antidiabetic studies of Co(II) glimepiride complex. International Journal of Research in Engineering and Innovation 2017; 1(3): 75-80.

[50] Wang X, Ling N, Zhang YW, Wang X, Yang HX. Synthesis, structure, antidiabetic and antioxidant properties of a new Co(II) complex with a flexible tripodal ligand. Inorg Chem Commun 2019; 110: 107585.
[http://dx.doi.org/10.1016/j.inoche.2019.107585]

[51] Mohamed M H, Entsar A S, Kholoud H R. Antidiabetic activity of cobalt–quercetin complex: A new potential candidate for diabetes treatment. J Appl Pharm Sci 2020; 10(12): 44-52.
[http://dx.doi.org/10.7324/JAPS.2020.101206]

[52] Sakthivel RV, Sankudevan P, Vennila P, Venkatesh G, Kaya S, Serdaroğlu G. Experimental and theoretical analysis of molecular structure, vibrational spectra and biological properties of the new Co(II), Ni(II) and Cu(II) Schiff base metal complexes. J Mol Struct 2021; 1233: 130097.
[http://dx.doi.org/10.1016/j.molstruc.2021.130097]

[53] Ajaz A, Shaheen MA, Ahmed M, *et al.* Synthesis of an amantadine-based novel Schiff base and its transition metal complexes as potential ALP, α-amylase, and α-glucosidase inhibitors. RSC Advances 2023; 13(5): 2756-67.
[http://dx.doi.org/10.1039/D2RA07051K] [PMID: 36756442]

[54] Venkataramaiah S, Venkatappa MM, Udagani C, Sannaningaiah D. Green-synthesized cobalt ferrite nanoparticle alleviated sodium nitrite-induced oxidative stress through its anti-oxidant property and displayed anti-inflammatory, anti-diabetic and anti-platelet activities. J Supercond Nov Magn 2024; 37(11-12): 1839-57.
[http://dx.doi.org/10.1007/s10948-024-06813-7]

CHAPTER 3

Recent Advances on Metal Dithiocarbamates and their Biological Applications

Vatsala Kumari[1], **Chote Lal Yadav**[2], **Anamika**[3] and **Krishna K. Manar**[1,*]

[1] *Department of Chemistry, Faculty of Science, University of Allahabad, Prayagraj, Uttar Pradesh, India*

[2] *Department of Chemistry and Environmental Science, Madan Mohan Malaviya University of Technology, Gorakhpur, Uttar Pradesh, India*

[3] *Department of Chemistry, Harcourt Butler Technical University, Kanpur, India*

Abstract: The growing demand for drug delivery systems and pharmaceuticals has attracted the interest of scientists worldwide, and the area of coordination/inorganic chemistry provides an enormous foundation. In this context, dithiocarbamate complexes have found many applications in biological systems. However, developing new dithiocarbamate complexes is necessary to advance cancer treatment, antiproliferative technologies, anti-leishmanial activity, and other biological activities. Recently, significant progress has been made in the development of dithiocarbamate complexes, in the consideration of the main mechanisms of action associated with these compounds, as well as in improving the main treatments for these diseases, which are being considered by the WHO (World Health Organization). Dithiocarbamate complexes are highly versatile ligands, known for their diverse structures and numerous coordination patterns that give rise to supramolecular networks. This diversity stems from the resonance structures the ligands can adopt in their complexes. A key factor driving the increasing attention in dithiocarbamate (dtc, ligand) chemistry is the ability to functionalize the different substituents on the N atom of the dtc ligand backbone. This modification can lead to intriguing crystal structures, enhanced materials, and potential biological applications. Overall, we provide a detailed review of dithiocarbamate complexes, their intriguing coordination activities, and their biological applications.

Keywords: Anti-leishmania, Anti-cancer, Biological activity, Dithiocarbamate complexes.

INTRODUCTION

The increasing interest in metal dithiocarbamate complexes arises from the ability to functionalize moieties on the dithiocarbamate (dtc) backbone, which can affect

* **Corresponding author Krishna K. Manar:** Department of Chemistry, Faculty of Science, University of Allahabad, Prayagraj, Uttar Pradesh, India; E-mails: kkmanar@allduniv.ac.in, krishnamanar@gmail.com

Abhay Nanda Srivastva (Ed.)
All rights reserved-© 2026 Bentham Science Publishers

fsteric and electronic behavior, thereby influencing their crystal structure and biological properties. Dithiocarbamates (dtc) are monoanionic chelating ligands that are readily synthesized from secondary (2°) amines. They form highly stable complexes with nearly all d-block and p-block elements [1 - 4]. Metal dithiocarbamate complexes have a wide range of applications across various fields, including agriculture, medicine, pharmaceuticals, manufacturing, catalysis, sensors, and materials science [5 - 12]. Numerous studies have been conducted to explore the properties of metal complexes formed with dithiocarbamate ligands. Research into the synthesis and properties of metallo-pharmaceuticals has surged since the discovery of the anticancer drug cisplatin and its analogs [13, 14]. Dithiocarbamates are easily synthesized from inexpensive and readily available secondary or primary amines by reacting with CS_2 in a basic solution, typically water, though methanol and other simple organic solvents can also be used. These ligands form stable complexes with nearly all transition metals, s-block elements, most p-block elements (excluding oxygen and the halogens), as well as the lanthanides and actinides [1 - 4]. The formation of dithiocarbamate complexes with various transition metals as well as main group metals is typically a straightforward process. It involves the addition of a water-soluble metal salt to an aqueous or methanolic solution containing a dithiocarbamate salt. This reaction leads to the formation and precipitation, which can be well collected through a simple filtration method, followed by washing and drying, resulting in a good to high yield of the dithio complexes. A key advantage of DTCs is their capacity to stabilize a large range of oxidation numbers across nearly all metals. This stabilization is ascribed to the resonance forms of dithiocarbamate (dtc) and thioureide forms (Fig. **1**). In its DTC form, classified as a soft ligand, the lone pair of electrons is localized on the sp^3 (tetrahedral) hybridized nitrogen atom, leading to a pyramidal shape of the substituents. On the other hand, the thioureide sp^2 hybridized resonance form, considered a hard ligand, adopts a planar structure, with the lone pair integrated into the C-N bond framework and extending toward the sulfur atom [15 - 17]. DTCs can coordinate with metals in various ways, with bidentate chelation being the dominant mode. This chelation significantly improves the thermodynamic stability of the compounds and increases their Coordination Number (C.N.) compared to simple monoanionic ligands.

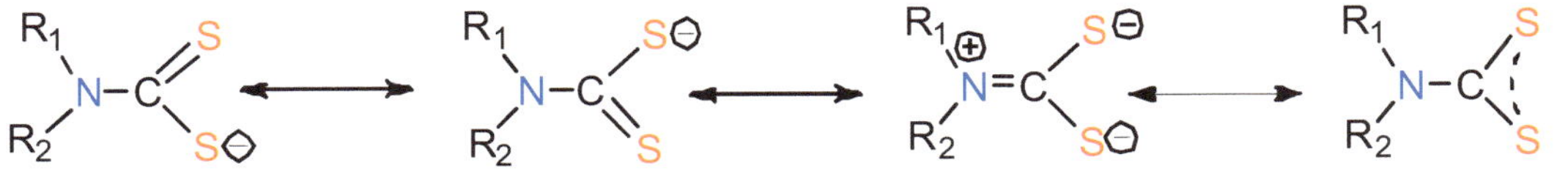

Fig. (1). Resonance form of dithiocarbamate and thiouride.

The small bite angle and the soft nature of the dithiocarbamate moiety allow these ligands to stabilize a broad range of transition metal ions and main group elements in various oxidation states [1 - 4, 18, 19]. The monodentate coordination mode is typically influenced by the absence of a vacant orbital on the metal with the appropriate symmetry to accommodate the lone pair of electrons from the second S atom. Additionally, the presence of other ligands already bonded to the metal can cause steric hindrance, preventing the coordination of the second S atom. This limitation may also occur due to the restricted vacancy left by a departing ligand, mainly in ligand exchange reactions. Dithiocarbamate ligands possess lipophilic properties and commonly coordinate with metals through various modes, with monodentate and anisobidentate being the most frequent. The electronic properties of these ligands can be adjusted by selecting different substituents [20]. Substituents on the nitrogen atom of the dithiocarbamate unit influence the overall complex structure, properties, and potential applications through secondary interactions [21]. Bulky substituents on the nitrogen atom typically hinder supramolecular aggregation by obstructing metal-sulfur interactions, while smaller substituents promote or enhance these interactions [22]. This, in turn, affects the stereo-configuration and electron density distribution, influencing the biological activity of various biomolecules, including proteins and enzymes, and potentially leading to toxic effects [23]. In this book chapter, we have summarized the synthesis of dithiocarbamate complexes and their medicinal applications such as anticancer activity, antileishmanial activity, effects on human breast cancer, anti-nematode activities, antibacterial activity, anticonvulsant properties, antianxiety activities, cytotoxic activity, DNA binding and cleavage, as well as antitumor activity.

General Synthesis of Dithiocarbamate

Dithiocarbamate (dtc) ligands, whether aliphatic or aromatic, are typically synthesized through the reaction of 1° or 2° amines with CS_2 and NaOH or KOH (Fig. **2**). In this process, nucleophiles (amines) first react directly with the electrophile (CS_2) to form a zwitterion. Subsequently, this zwitterion reacts with base (sodium or potassium hydroxide) to generate monoanionic dithiocarbamate ligands and water. The reaction is generally conducted under exothermic conditions, starting at low temperatures (0 °C or in an ice bath) and gradually increasing to room temperature [1, 11, 15, 24 - 29].

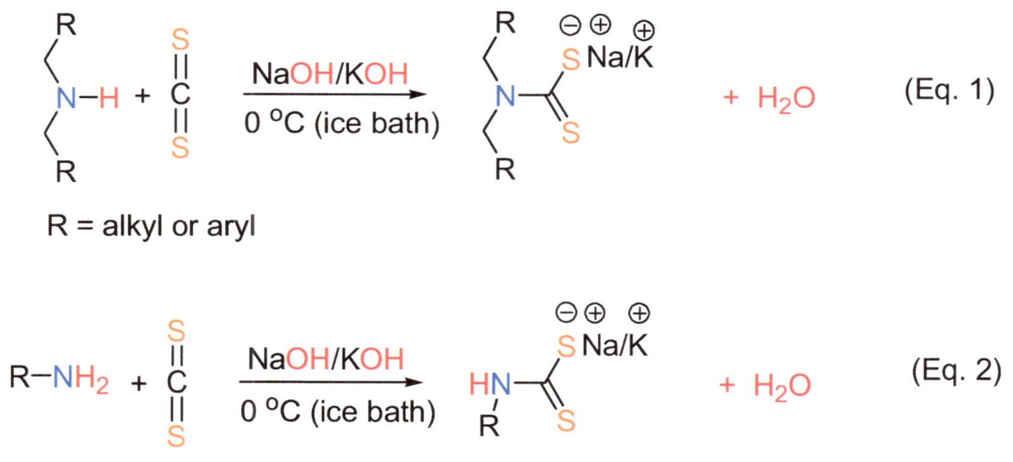

Fig. (2). A line drawing that illustrates the general synthetic route of monoanionic dithiocarbamate (dtc) ligands.

Synthesis, Characterizations, and Biological Applications of the Dithiocarbamate Complexes

Bharty and group reported the synthesis and structural analysis of metal complexes containing azepane-1-carbodithioate, with the aim of targeting human breast cancer [30]. Their study concentrated on the cytotoxic activity of these synthesized complexes. The team successfully synthesized complexes (**1-4**) using potassium azepane-1-carbodithioate (acdt) as the ligand (Scheme **1**).

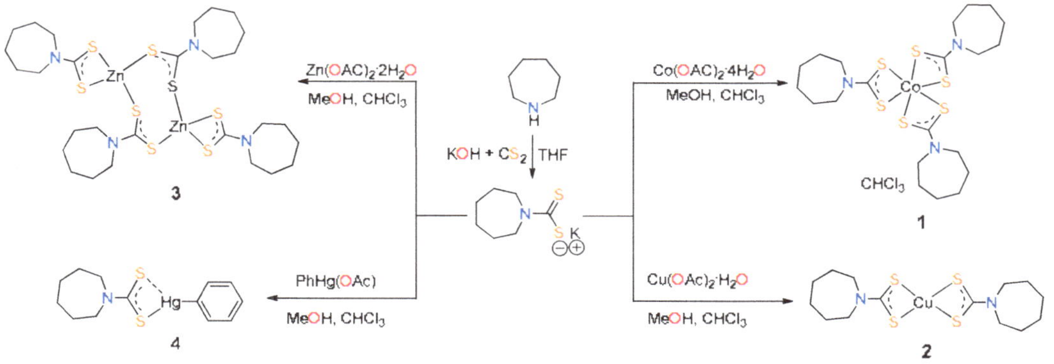

Scheme 1. General synthesis route of complexes **1-4**.

The crystal structure of the complexes was determined using Single-Crystal X-Ray Diffraction (SCXRD). In addition, they tested the cytotoxic activity of both the ligand and the complexes against the human breast cancer (MDA-MB-231) cell line. The results showed that complex **1** exhibited the maximum cytotoxicity compared to the other complexes. Furthermore, they investigated the glucose metabolism in the MDA-MB-231 cell line. By inhibiting glucose uptake, cancer cells may experience starvation, potentially leading to cell loss due to energy depletion. The results showed that the group treated with complex **1** exhibited significantly better inhibition of glucose uptake compared to those treated with the free ligand and complexes. Additionally, the team measured extracellular nitric oxide (NO) levels in MDA-MB-231 cells treated with the free ligand and its complexes **1-4**. The result indicated that the ligand and its complexes stimulated nitric oxide production in the cancer cells. The Co(III) complex **1** notably increased the NO level in cells (MDA-MB-231) to 75.91 μM, while complexes **2-4** produced NO levels of 62.91 (**2**), 49.24 (**3**), and 33.10 (**4**) μM. In conclusion, complex **1** showed significant anticancer activity against the MDA-MB-231 cell line. This indicates that Co(III) complex **1** not only stimulates nitric oxide production but also enhances oxidative stress in MDA-MB-231 cells more effectively than the free ligand and other complexes.

Ajibade and his team have published the synthesis, characterization, and *in vitro* anticancer screening of transition metal methoxyphenyl dithiocarbamato complexes, including Mn(II) **5**, Co(II) **6**, Cu(II) **7**, Zn(II) **8**, and Pt(II) **9** ions [31]. They synthesized these transition metal complexes with methoxyphenyl dithiocarbamate (dtc) ligands (Scheme **2**) and fully characterized them by various spectroscopic techniques. The characterization (spectroscopic) data revealed symmetrical bidentate chelating of the dtc ligand to the transition metal ions (Co(II), Cu(II), Zn(II), and Pt(II). For *in vitro* testing, the anticancer activity of the synthesized complexes was evaluated against HeLa cells (human fetal lung cancer cells) and their normal cell line. All complexes, except complex **7**, showed cytotoxic effects on HeLa cells in the first round of screening. When tested in the cancer cell line, the cobalt(II) and zinc(II) complexes emerged as the most active. Zinc(II) complex displayed the highest cancer cell selectivity, with an SI of 2.0, while cobalt(II) complex had an SI of 1.4. In contrast, the manganese(II) complex showed no cancer cell selectivity, and the platinum(II) complex was highly cytotoxic toward normal cells, resulting in an SI of 0.2. The anticancer activity of these synthesized metal complexes suggests they are promising candidates for the development of inorganic anticancer agents.

Andrew and his team conducted a study on the synthesis, characterization, and *in vitro* anticancer properties of chlorido(triphenylphosphine)ruthenium(II) dithiocarbamate complexes [32]. They synthesized three ruthenium(II) dithiocarbamate complexes **10-12** (Scheme **3**). The characterization of these complexes was performed through elemental analysis and various spectroscopic methods.

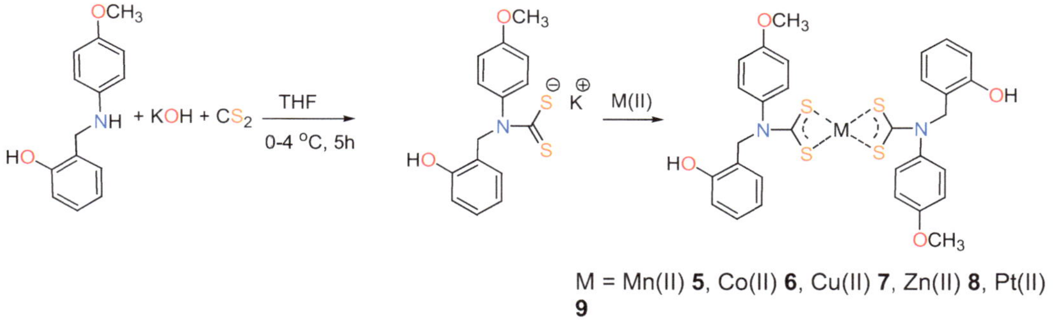

Scheme 2. Line drawing of the synthesis of complexes **5-9**.

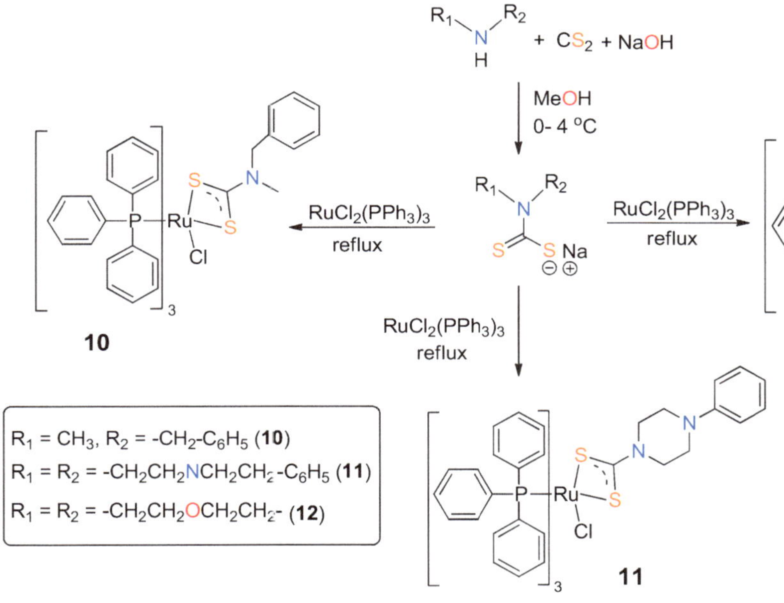

Scheme 3. Line drawing of Ru(II) dithiocarbamate complexes **10-12**.

In these complexes, the ruthenium(II) ion adopts an octahedral geometry, with one dithiocarbamato anion serving as a bidentate chelating ligand. The anticancer activity of these complexes was evaluated using the sulforhodamine B assay on various human cancer cell lines, including renal, melanoma, and breast cancers. Complex **10** showed the highest cytotoxicity against breast cancer cells, while complex **12** exhibited the lowest activity against renal cells.

Nami and their group have reported the synthesis and spectral characterization of organotin(IV) complexes with the isophthaloylbis(piperazine-4,1-diyl)bis(methylene)dicarbamodithioate ligand, while also screening their anticancer and anti-nematode activities [33]. The organo Sn(IV) complexes (**13-15**) were synthesized (Scheme **4**) and thoroughly analyzed by various spectroscopic methods, including ^{119}Sn NMR, as well as thermal analysis. IR spectra indicated symmetrical coordination of the ligand in all the complexes. Additionally, ^{13}C NMR and ^{119}Sn NMR confirmed six-coordinate geometry around the metal center. Notably, ligand and its complexes were evaluated for anti-breast cancer and anti-nematode activities. Among them, complexes **14** and **15** exhibited more effective anti-breast cancer activity compared to the other complexes. Additionally, all synthesized complexes demonstrated anti-nematode activity.

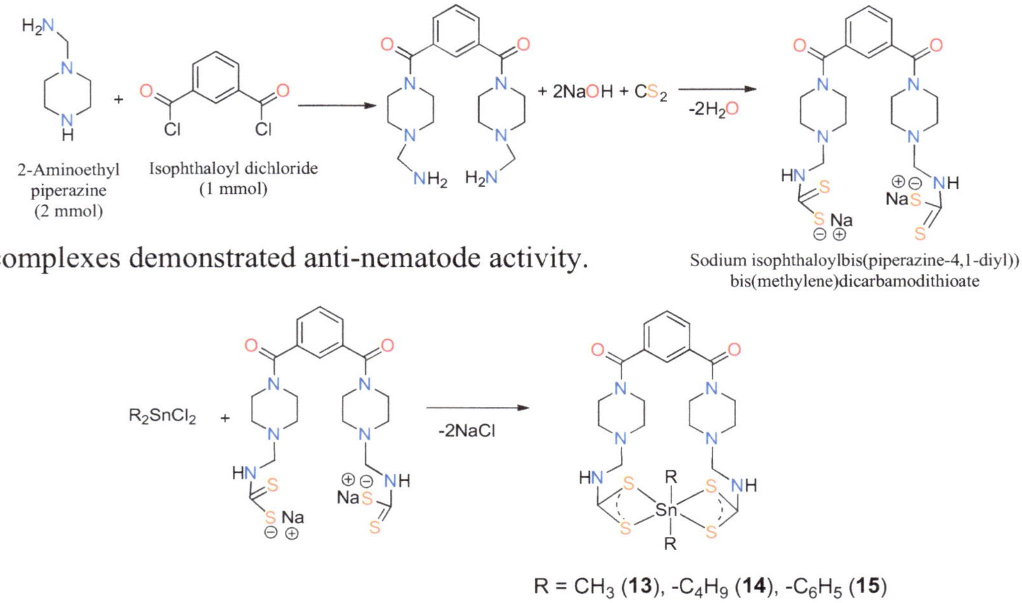

complexes demonstrated anti-nematode activity.

Scheme 4. A line drawing illustrating the synthesis of the dithiocarbamate ligand and its corresponding complexes **13-15**.

Ajibade and his team reported their findings on the synthesis and anticancer activity of dithiocarbamate complexes**16-19** (Scheme **5**). They also determined the crystal structures of the Cu(II) and Pt(II) complexes (**17** and **19**) [34]. The synthesized complexes were thoroughly analyzed by spectroscopic techniques, and their crystal structures were determined through SCXRD analysis. The *in vitro* cytotoxic activity of the complexes was evaluated against HeLa cell lines as well as the normal cell lines MRC5 and MRC5-SV2. Notably, the complexes exhibited significant toxicity against various cancer cell lines. Among them, copper (II) and zinc (II) complexes demonstrated the highest efficiency, showing nearly identical cytotoxicity against both cancer cell lines. In contrast, only the zinc(II) complexes demonstrated greater selectivity for cancer cells over normal cells. However, the Mn(II) complexes were the least effective. In contrast, Pt(II) complexes exhibited the lowest selectivity for cancer cells, inflicting significantly greater toxicity on the MRC-5 cells than on its cancerous counterpart, MRC5-SV2. Thus, these synthesized complexes could pave the way for the creation of innovative inorganic drug leads, potentially leading to the development of novel anti-cancer therapies. These therapies could be delivered directly to tumor tissues or *via* specialized delivery systems.

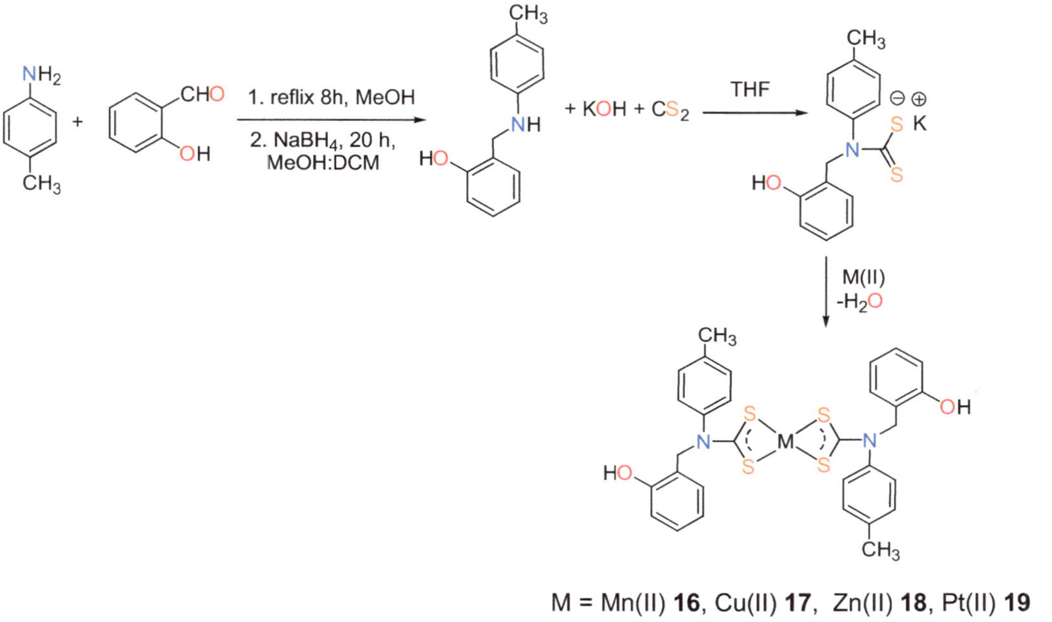

Scheme 5. Line drawing of the synthesis of complexes**16-19**.

Prasad and his team published their research on the synthesis, characterization, and *in vitro* antibacterial and cytotoxic activities of Co(II) (**20**), Ni(II) (**21**), Cu(II) (**22**), and Zn(II) (**23**) complexes, which feature the N-(4-methoxybenzyl-N-(phenylethyl) dithiocarbamate ligand [35]. They synthesized the cobalt(II) (**20**), nickel(II) (**21**), copper(II) (**22**), and zinc(II) (**23**) complexes (Scheme **6**) using the mentioned ligand, characterizing them through elemental analysis (EA) and various spectroscopic methods. Additionally, they examined the luminescent behavior of the synthesized complexes and ligands. Notably, the complexes displayed weak luminescence compared to the ligands, which was attributed to metal-perturbed MLCT states. The crystal structure of the Zn(II) complex (**23**) was elucidated by SCXRD, revealing a distorted square pyramidal geometry around the Zn(II) metal ion. The UV-Vis and luminescent properties of both the DTC ligand and the complexes were measured in CH_3OH solution. The ligand exhibited an absorption band at 301 nm, while the complexes showed absorption bands between 265-275 nm, indicative of intra-ligand charge transfer (ILCT) transitions. Complexes **20-22** displayed absorption transitions between 325-445 nm, corresponding to LMCT transitions. Antibacterial activity was evaluated for all complexes against *Staphylococcus aureus, Bacillus subtilis*, *Clostridium sporogenes*, and *Escherichia coli*. The results revealed that complexes **21** and **23** exhibited significant antibacterial activity compared to the others. Additionally, the cytotoxicity study against breast cancer showed that complex **21** displayed promising anticancer properties.

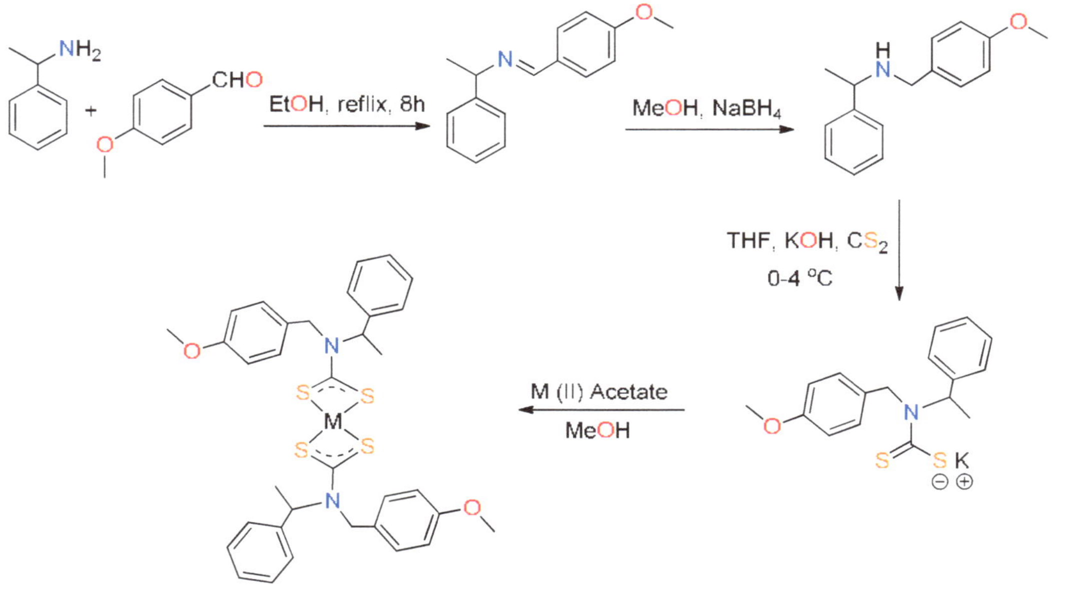

M = Co(II) **20**, Ni(II) **21**, Cu(II) **22**, Zn(II) **23**

Scheme 6. Schematic illustration of the synthesis of the ligand and transition metal dithiocarbamate complexes**20-23**.

Prasad and his team also reported the synthesis, crystal structure, and evaluation of the anticonvulsant and antianxiety activities of three novel dithiocarbamate complexes of Ni(II) (**24**), Cu(II) (**25**), and Zn(II) (**26**) (Scheme **7**) [36]. The complexes were fully characterized using Elemental Analysis (EA), magnetic susceptibility, and various spectroscopic methods. Single-crystal X-ray diffraction analysis of complexes **25** and **26** revealed that the zinc (II) complex adopts a tetrahedral geometry, while the nickel(II) and copper(II) complexes have a square planar geometry around the metal center. The anticonvulsant potential of the ligand fbnm and its nickel(II) (**24**), copper(II) (**25**), and zinc(II) (**26**) complexes was assessed through the MES and PTZ-induced convulsion tests, suggesting that complex **25** might be a promising anticonvulsant candidate, subject to further biochemical or neurotransmitter studies to manage generalized tonic-clonic seizures. On the other hand, the evaluation of antianxiety activity using the Elevated Plus Maze and Open Field tests showed that none of the synthesized complexes exhibited significant anti-anxiety effects.

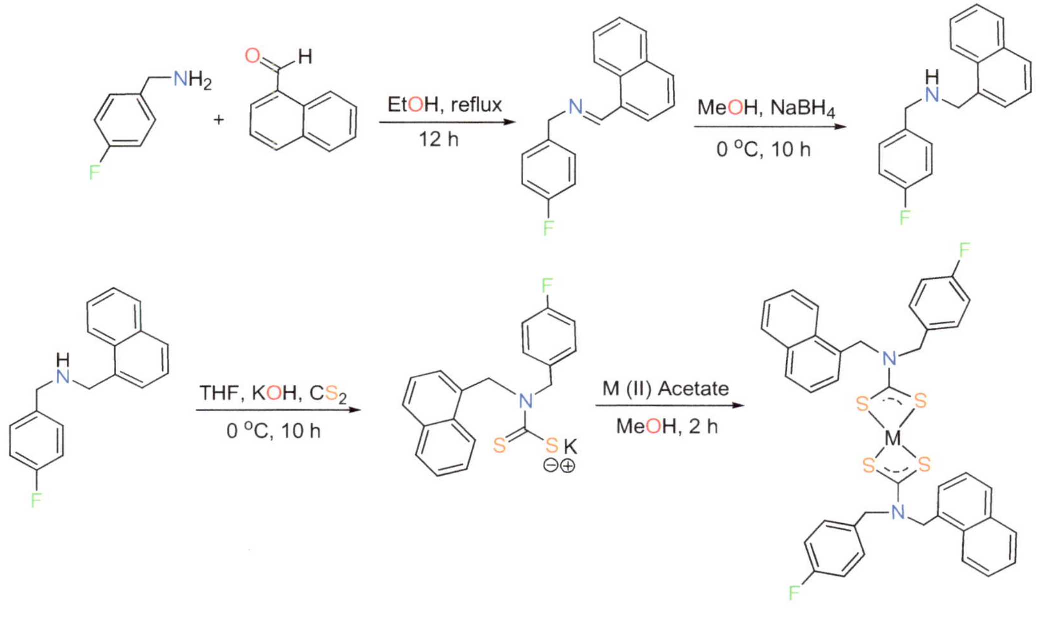

M = Ni(II) **24**, Cu(II) **25**, Zn(II) **26**

Scheme 7. Schematic illustration of the synthesis of dithiocarbamate complexes **24-26**.

In a study by Moghadam and the group, platinum(II) complexes with tertamyl dithiocarbamate ligands were studied for their anticancer activity, DNA targeting activities, molecular dynamics, and biological activity [37]. They have reported the design and synthesis of several anti-tumor platinum complexes incorporating dithiocarbamate ligands. Biological evaluations of these complexes have also been carried out (Fig. **3**) [38, 39].

Furthermore, they synthesized the platinum(II) compound [Pt(phen)(tertamyl.dtc)]NO_3 (**27**) (Fig. **4**), and characterized it using various spectroscopic techniques. Various spectroscopic techniques and simulation studies were employed to investigate the DNA-binding activity of these platinum dithiocarbamate derivatives.

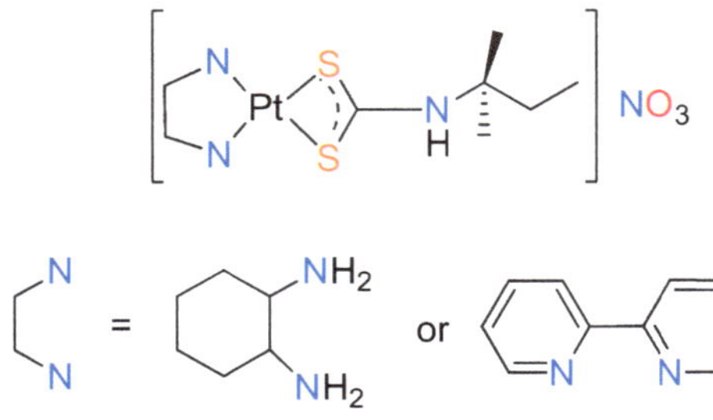

Fig. (3). Tertamyl dithiocarbamate Pt(II) complexes.

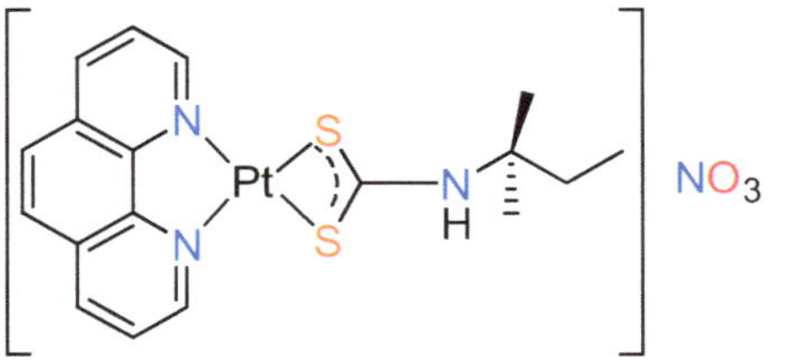

27

Fig. (4). Molecular structure of complex **27**.

Through UV-Vis and thermal stability studies, they assessed the degree of groove binding between the platinum complex and CT-DNA. Based on the findings and the simulated results, the platinum complex can bind to DNA and form stable DNA complexes by binding to the minor grooves on G and T centers. They have screened the cytotoxic activity of the synthesized platinum complex and oxaliplatin against several human cell lines, including MCF7, HCT116, HT29, and Panc1, at concentrations ranging from 0 to 800 μM. It was found that the synthesized complex has a lower IC_{50} value against HT29 and Panc1 cells as compared to oxaliplatin. As a result, the synthesized Pt(II) complex had a greater efficacy than oxaliplatin. They have observed that the N, N donor ligands in the complex structure can affect the potential drug in the following order: dach<bpy<phen. Therefore, the complex containing more conjugated phen was more biologically active.

Thirumaran and colleagues reported the Pd(II) **28** and Pt(II) **29** complexes, along with their anticancer studies [40]. They outlined the general synthetic method for the dithiocarbamate ligand and its corresponding metal complexes (Scheme **8**).

The synthesized complexes were characterized through Elemental Analysis (EA) and various spectroscopic methods. Notably, the *in vitro* anticancer activity of the palladium(II) complex **28** and platinum(II) complex **29** was evaluated against the MCF-7 cell line using the MTT assay. The platinum complex **29** exhibited better activity compared to the palladium(II) complex **28**. IC_{50} values and cell viability data were determined, showing that both complexes demonstrated anticancer activity in a dose-dependent manner. Furthermore, complex **29** had a lower IC_{50} value (32 μM) compared to complex **28** (60 μM), indicating that complex **29** was more effective against MCF-7 cells.

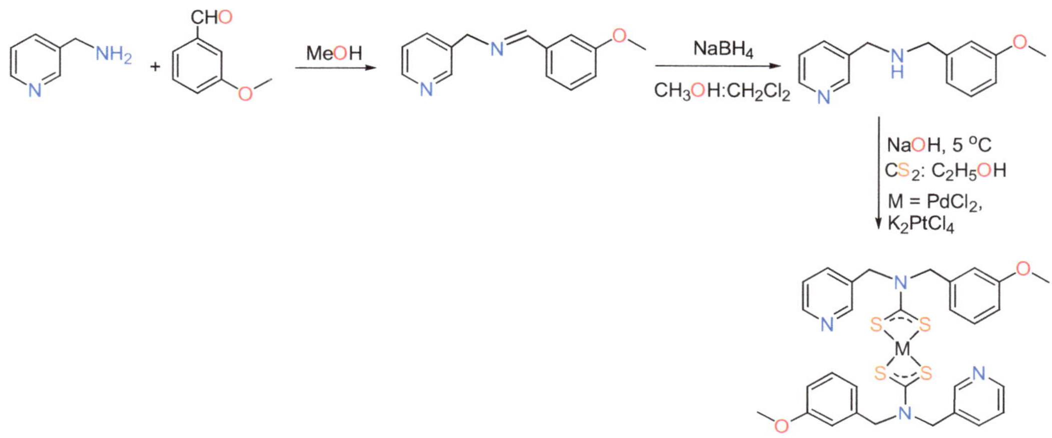

M = Pd(II) **28**), Pt(II) **29**

Scheme 8. Line drawing of the synthesis of dithiocarbamate ligand and respective metal complexes **28-29.**

Singh and his group published a study exploring the effect of functional groups on the structures of Zn(II) dithiocarbamates, with a focus on anti-leishmanial activity [41]. They synthesized a series of Zn(II) dithiocarbamate complexes **30-35** using functionalized dithiocarbamate ligands. The complexes were characterized through elemental analysis (EA) and various spectroscopic methods. Structural determination by SCXRD revealed that the complexes exhibited distorted tetrahedral (Td) and trigonal bipyramidal/square pyramidal (TBP/SP) coordination geometries. The synthesized complexes included dimeric structures for complexes **30** and **31**, while complexes **32**, **33**, and **34** displayed distorted tetrahedral and TBP/SP coordination, respectively. In contrast, complex **35** was mononuclear and exhibited a distorted tetrahedral geometry. Anti-leishmanial testing demonstrated that complexes **33** and **35** (Fig. **5**) exhibited significant antipromastigote and anti-amastigote activities, respectively. Cytotoxicity assays showed that both compounds were toxic to promastigotes but exhibited lower toxicity against the RAW 264.7 cell line at varying concentrations.

Singh and his team have investigated the effect of functional groups on the structures and anti-leishmanial properties of novel Bi(III) dithiocarbamate complexes [42]. They synthesized six new functionalized Bi(III) dithiocarbamate complexes **36-41**. The complexes were analyzed by Elemental Analysis (EA),

Powder X-Ray Diffraction (PXRD), and various spectroscopic methods, with their structures elucidated *via* SCXRD. Anti-leishmanial testing showed that compounds **40** and **41** (Fig. **6**) exhibited promising anti-promastigote and anti-amastigote activities. Cytotoxicity assays demonstrated that all the compounds were toxic to promastigotes while showing lower toxicity toward the RAW 264.7 cell line at different concentrations.

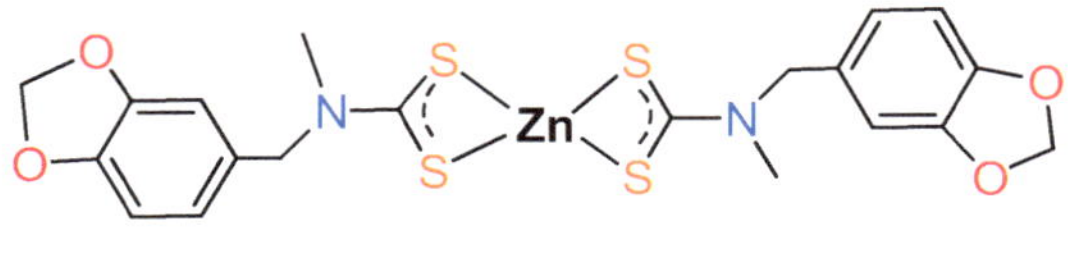

33

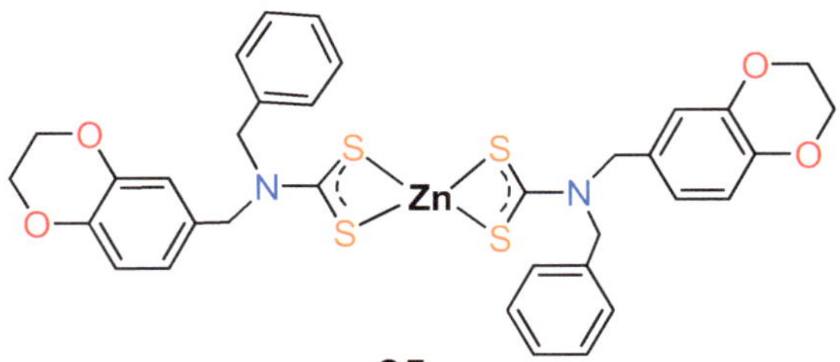

35

Fig. (5). Line drawing of the Zn(II) dithiocarbamate complexes **33** and **35**.

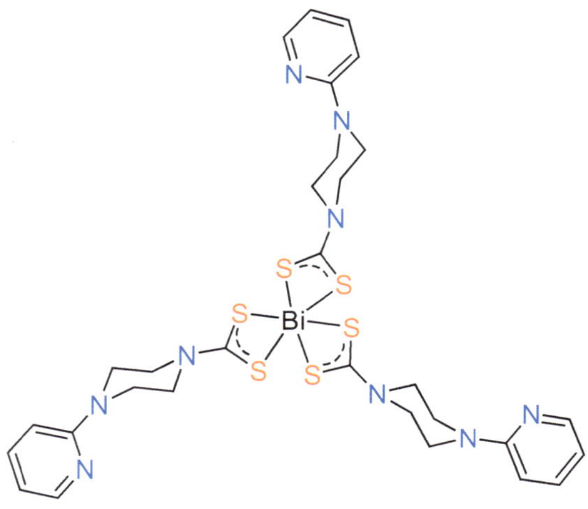

40

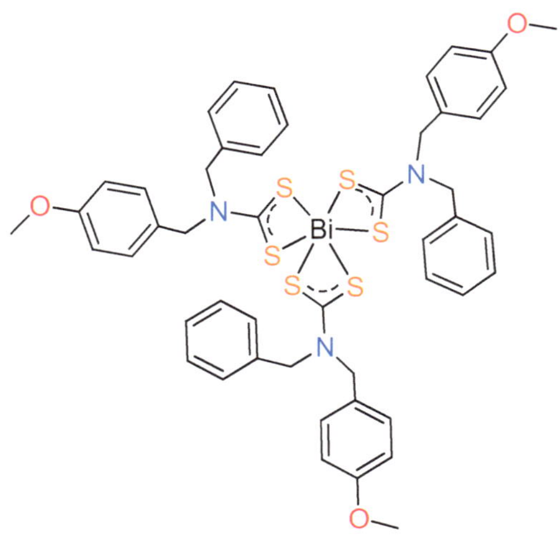

41

Fig. (6). Line drawing the Bi(III) dithiocarbamate complexes **40 and 41**.

Singh and colleagues have reported on the synthesis of nickel (II), palladium (II), and platinum (II) β-oxodithioester complexes and their anti-leishmanial properties [43]. They synthesized a series of functionalized planar β-oxodithioester cis-chelate complexes **42-53** (Scheme **9**). The complexes were characterized using elemental analysis (EA) and various spectroscopic methods. SCXRD was employed to determine the crystal structures of complexes **43–45**, **49**, and **52**. The anti-leishmanial potential of these complexes was evaluated, revealing that complexes **48** and **50** demonstrated significant anti-promastigote and anti-amastigote activities (Fig. **7**). Cytotoxicity assays showed that both complexes were toxic to promastigotes but exhibited reduced toxicity toward RAW 264.7 cell lines at different concentrations.

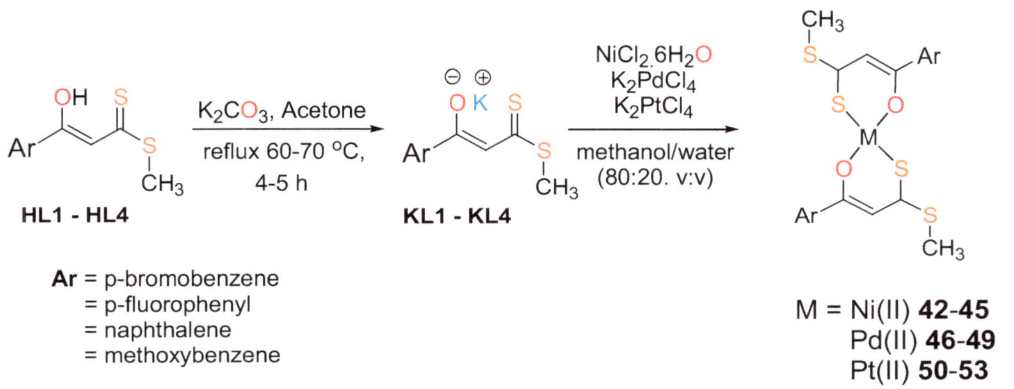

Scheme 9. Generalized synthesis of β-oxodithioester ligands and complexes **42-53**.

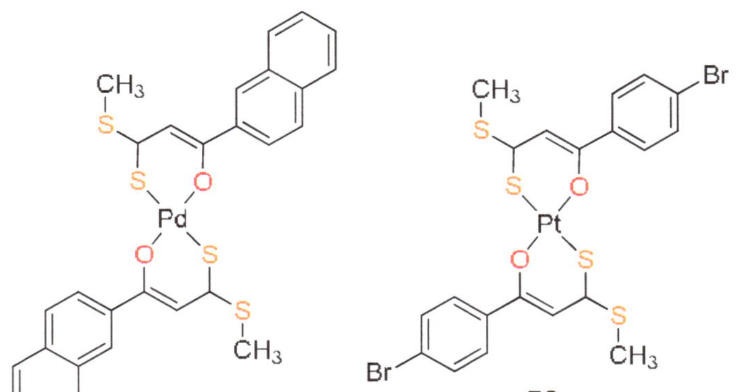

Fig. (7). β-oxodithioester complexes of **48** and **50**.

Furthermore, Singh and colleagues have reported on the synthesis, anticancer activity, and DNA-binding/cleavage studies of planar trans-copper(II) β-oxodithioester complexes [44]. They synthesized copper (II) β-oxodithioester complexes **54-58** (Scheme **10**). The complexes were analyzed by Elemental Analysis (EA) and various spectroscopic techniques. The structures of the ligand (HL3) and its corresponding complex (**56**) were determined by single-crystal X-ray diffraction (SCXRD). All copper (II) complexes demonstrated effective DNA binding and DNA cleavage in a concentration-dependent manner, following the order: **54**>**56**>**58**>**55**>**57**. Among these, the pyridyl- and methoxy-functionalized complexes **54** and **56** exhibited enhanced self-activating abilities in DNA cleavage. Furthermore, the compounds showed notable variation in IC_{50} values against the MCF-7 cell line. Treatment with these complexes resulted in a dose-dependent increase in apoptotic cell death in the RAW 264.7 cell line.

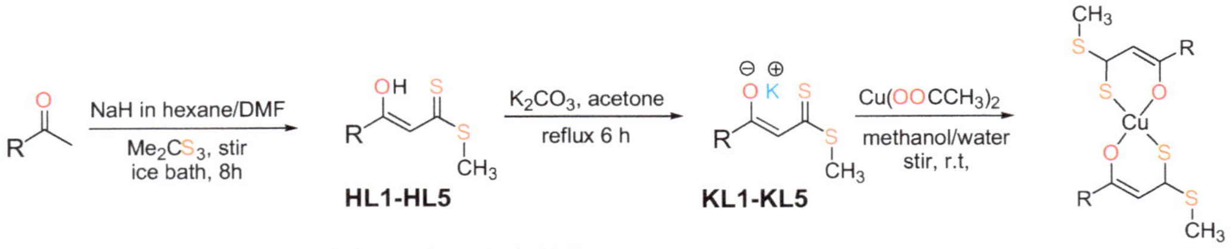

54-58

Scheme 10. Generalized synthetic route of β-oxodithioester ligands and Cu(II) β-oxodithioester complexes **54-58**.

Singh *et al.* investigated the zinc (II) β-oxodithioester compounds (**59**, **60**, Scheme **11**), along with their DNA binding and cleavage activities [45].

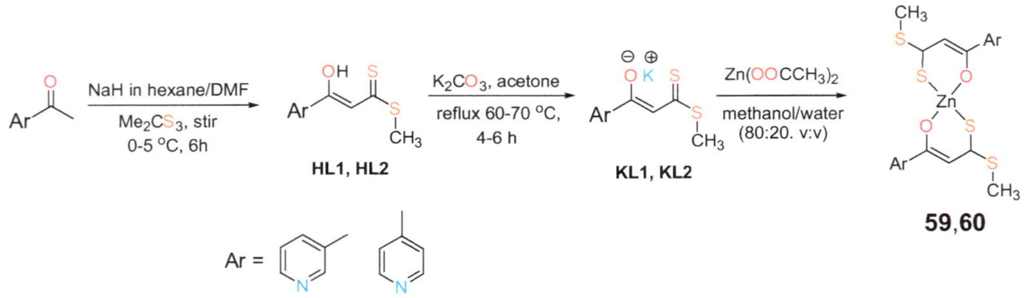

59,60

Scheme 11. Generalized synthetic route of β-oxodithioester ligands and Zn(II) β-oxodithioester complexes **59** and **60**.

The complexes were analyzed by Elemental Analysis (EA), ESI-MS, and various spectroscopic methods. The structure of compound **59** was elucidated by SCXRD. DNA binding and cleavage studies revealed that complex **59**, which contains nitrogen in the 3-position of the pyridine backbone, exhibited enhanced DNA cleavage activity in the presence of ascorbic acid. However, it showed protective effects against DNA cleavage when exposed to H_2O_2 or DMSO. Additionally, compound **59** demonstrated cytotoxicity against the MCF-7. The combination of notable cytotoxic activity and DNA cleavage ability in the presence of ascorbic acid suggests its potential as an anticancer agent.

Singh and colleagues reported the synthesis, crystal structures, and properties of homoleptic Ni(II) and Pd(II) β-oxodithioester chelates [46]. They synthesized six new cis-chelate complexes, **61-66** (Scheme **12**). The complexes were analyzed by elemental analysis (EA) and various spectroscopic methods. The crystal structures of complexes**62-66** were determined through SCXRD. Antileishmanial studies showed that complexes **62**, **63**, and **66** exhibited significant antipromastigote and anti-amastigote activities (Fig. **8**). Cytotoxicity assays revealed that these complexes were toxic to promastigotes but demonstrated lower toxicity against RAW 264.7 cell lines at various concentrations.

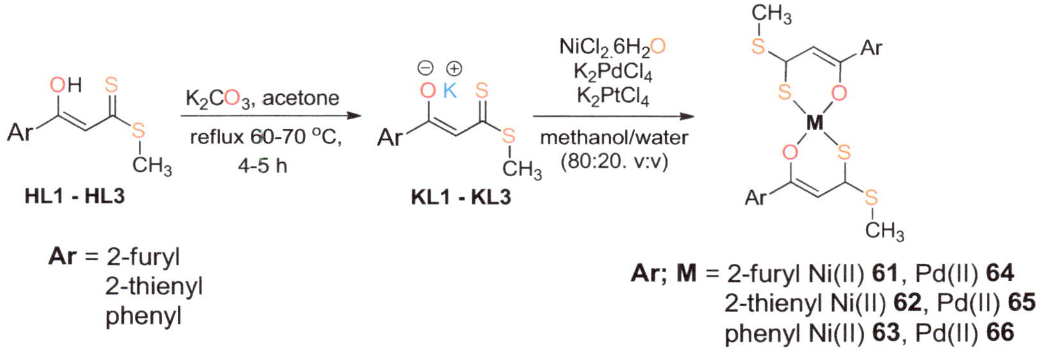

Scheme 12. General synthetic route of the complexes **61-66**.

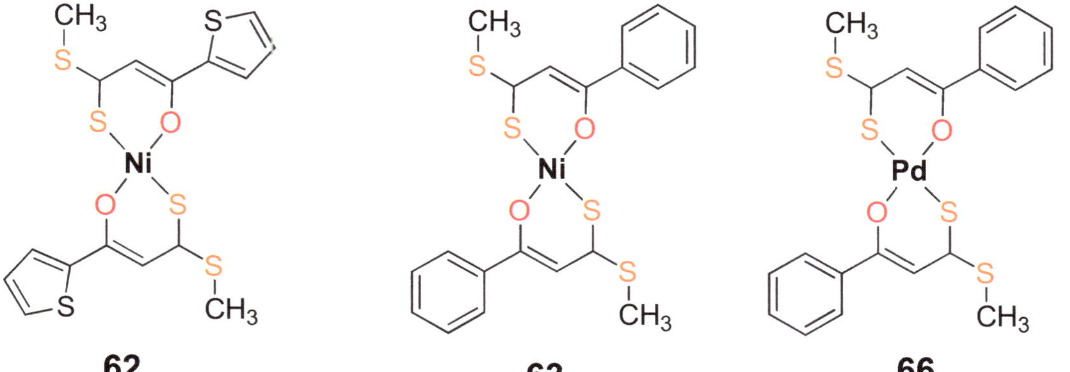

Fig. (8). Complexes **62, 63,** and **66** are active against antileishmanial activity.

Paca and colleagues reported the synthesis, single crystal X-ray diffraction (SCXRD), and anticancer activity of palladium (II) dithiocarbamate complexes

[47]. They synthesized the complexes [Pd(L)$_2$], where L includes bis(piperidinyl-dithiocarbamato) (L1, **67**), bis(1-phenylpiperazinyldithiocarbamato) (L2, **68**), bis(phenyldithiocarbamato) (L3, **69**), and bis(4-benzylpiperidinyl) (L4, **70**) (Fig. **9**).

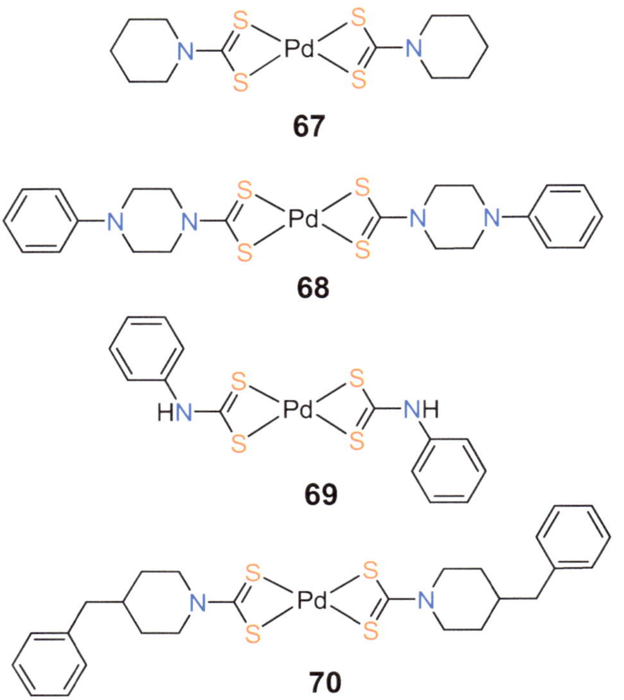

Fig. (9). Line drawing of complexes **67-70**.

The complexes were fully analyzed by Elemental Analysis (EA) and various spectroscopic methods, with their structures elucidated through SCXRD. Cytotoxicity studies on three human cancer cell lines revealed that all complexes exhibited high to moderate activity. Complexes **67** and **69** exhibited strong activity against all three cell lines, surpassing the efficacy of the consequent dithiocarbamate (dtc) ligands. In contrast, the anticancer potency of complexes **68** and **70** was generally higher than that of the respective palladium (II) complexes.

Prasad and colleagues reported the Ni(II), Cu(II), and Zn(II) complexes of N-bis(4-methoxybenzyl) dithiocarbamate and evaluated their antitumor activity [48]. The synthesized complexes are depicted in Scheme **13**. These complexes were analyzed by elemental analysis (EA) and various spectroscopic methods. The structure of complex **73** was elucidated by SCXRD. Antitumor studies revealed significant activity against Dalton's lymphoma cells in a concentration-dependent manner, while exhibiting relatively low toxicity toward normal human cells, including lymphocytes and monocytes.

Ajibade and colleagues reported the copper(II) complex **74** and zinc(II) complex **75** of morpholinyldithiocarbamato and their anticancer studies [49]. The complexes were synthesized as $[Zn(MphDTC)_2]_2$, **74** and $[Cu(MphDTC)_2]$ **75**, where MphDTC stands for morpholinyldithiocarbamate (Scheme **14**). Both complexes were analyzed by elemental analysis (EA) and various spectroscopic methods, with structural elucidation achieved through single-crystal X-ray diffraction (SCXRD). The Cu(II) complex was identified as a mononuclear complex, with the copper (II) ion coordinated to two morpholinyldithiocarbamate ligands, resulting in a four-coordinate distorted square planar geometry. In contrast, the zinc (II) complex was found to be a dimer, with each zinc ion coordinated to two bidentate dithiocarbamate ligands, where one ligand acts as a chelating agent and the other as a bridging ligand. The anticancer activities of the dithiocarbamate ligand and its copper(II) and zinc(II) complexes were assessed against melanoma, renal, and breast cancer cell lines using a Sulforhodamine B assay. The dithiocarbamate ligand exhibited greater activity than the standard drug parthenolide against renal and breast cancer cell lines, while the Zn(II) complex demonstrated the highest activity against breast cancer. The Cu(II) complex demonstrated similar effectiveness to the standard in targeting renal and breast cancer cell lines, but it displayed enhanced potency against melanoma when compared to parthenolide.

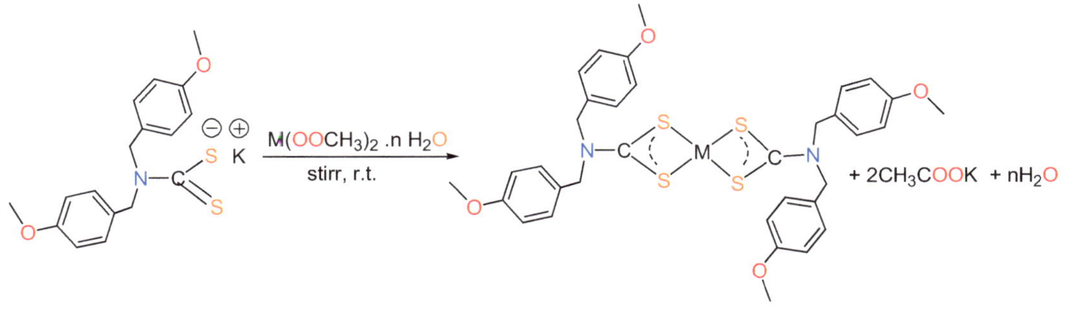

M = Ni(II) **71**, Cu(II) **72**, Zn(II) **73**

Scheme 13. General synthetic pathway for complexes **71-73**.

Ajibade and colleagues reported the copper(II) complex **76**, zinc(II) complex **77**, and platinum(II) complex **78** of bis(1-phenylpiperazine dithiocarbamate) and their anticancer studies [50]. The complexes were synthesized (Scheme **15**) and analyzed by Elemental Analysis (EA) and various spectroscopic methods. Crystal structure was performed for the zinc(II) and platinum(II) complexes. The structure of the zinc(II) complex reveals a centrosymmetric dimeric molecule, where both zinc ions are coordinated to two ligand molecules, acting either as bidentate chelating agents or as bridging ligands. The coordination environment of both Zn ions shows a distorted tetrahedral geometry. In contrast, the platinum(II) complex exists as a monomeric unit, where the platinum(II) ion is coordinated to four sulfur atoms from two dtc ligands, resulting in a slightly distorted square planar geometry. The anticancer activity of these compounds was evaluated against three cancer cell lines, with the platinum(II) complex demonstrating an activity greater than 100 μM in all tested cell lines.

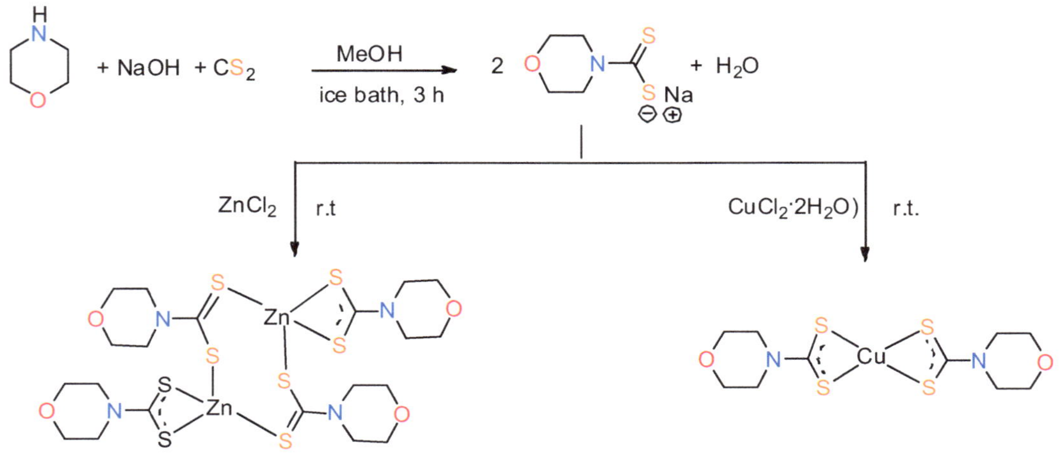

74 **75**

Scheme 14. General synthetic pathway for complexes **74** and **75**.

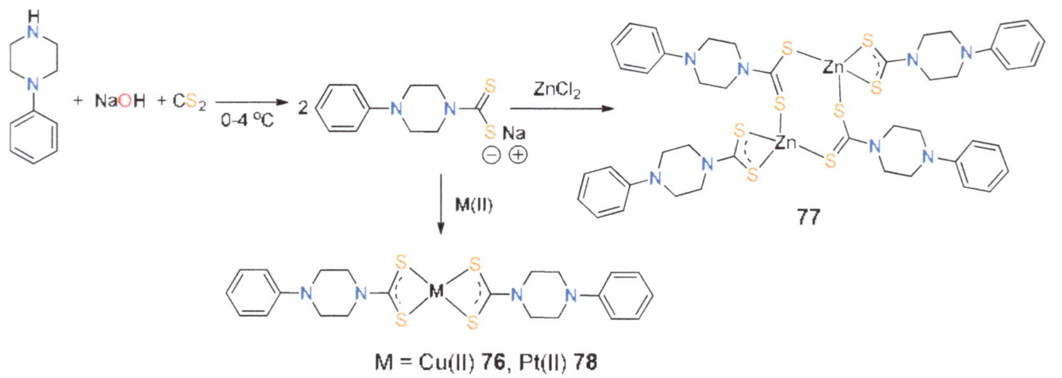

Scheme 15. Synthesis of dithiocarbamate ligand and complexes **76-78**.

Ajibade and colleagues reported the synthesis of Cu(II), Zn(II), and Pt(II) bis-(N-methyl-1-phenyldithiocarbamato) complexes (complexes **79-81**) and evaluated their anticancer activity (Scheme **16**) [51]. They synthesized the N-methyl-

1-phenyldithiocarbamate complexes **79-81**, which were analyzed by Elemental Analysis (EA) and spectroscopic methods. The electronic spectra of the platinum(II) and copper(II) complexes indicate a square planar geometry. The crystal structure of [Cu(N-mpDTC)$_2$] established a square planar structural arrangement. *In vitro* antiproliferative assays on three human cancer cell lines showed that the Zn complex demonstrated greater activity than the copper and platinum complexes.

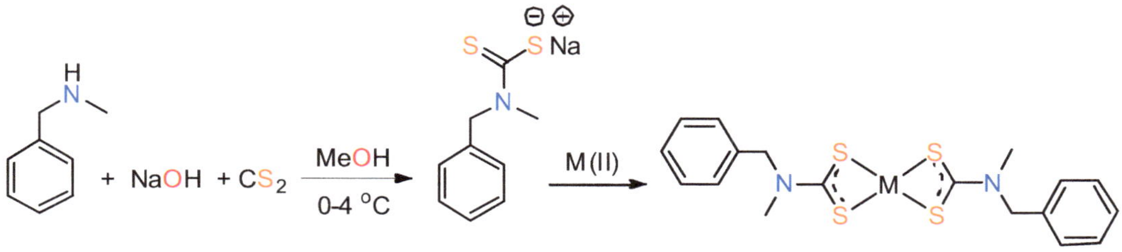

Scheme 16. General synthetic route of the complexes **79-81**.

Ajibade and colleagues studied the photocatalytic degradation of organic dyes, and assessed the anticancer activity of CuS nanoparticles derived from the bis(diallyl dithiocarbamato)copper(II) complex **82** [52]. The complex was synthesized and characterized using SCXRD (Fig. **10**), revealing a dimeric structure with two Cu(II) ions, each coordinated by two diallyldithiocarbamato bidentate chelating ligands. The centrosymmetrically related diallyldithiocarbamato ligands bridge the sulfur atoms of neighboring chelating ligands, resulting in a square pyramidal (SP) geometry around the Cu(II) metal ions. CuS nanoparticles were synthesized at 160 °C *via* the thermolysis of the Cu(II) complex in dodecylamine, hexadecylamine, and octadecylamine to investigate how capping agents influence the structural characteristics, photocatalytic, and anticancer activities of the resulting nanoparticles. HRTEM images showed spherical particles ranging from 2 to 25 nm in size. The copper sulfide nanophotocatalysts exhibited high efficiency in degrading bromophenol blue, with CuS-DDA achieving the highest photocatalytic efficiency. In terms of anticancer activity, CuS-HDA showed the highest potency, with IC_{50} values of 1.22 μg/mL against HEK293 cells and 28.78 μg/mL against HeLa cells.

Rehman and colleagues have published a study on the synthesis, characterization, and anticancer activity of monofunctional Pt(II) dithiocarbamate complexes [53]. The heteroleptic Pt(II) complexes **83-85** are shown in Scheme **17**. The ligands L consist of tri(4-fluorophenylphosphine) for compounds **83** and **85**, and tri(4-chlorophenylphosphine) for complex **84**. These complexes have been synthesized and analyzed by different analytical methods. Structural analysis reveals that these complexes adopt a square planar geometry, with steric hindrance from the aromatic C–H groups of the phosphine ligands, as confirmed by SCXRD. In complex **83**, the platinum square plane is blocked by two axially oriented H atoms, whereas complexes**84** and **85** are hindered by only one. DNA-binding studies using UV-Vis spectroscopy showed that complex **83** exhibits a stronger electrostatic interaction than complexes**84** and **85**, a result further supported by viscometry and Cyclic Voltammetry (CV). *In vitro* anticancer assays using the MTT method demonstrated the strong potency of these complexes against five distinct cancer cell lines. The enhanced activity of complex **83** relative to complexes**84** and **85** aligns with its DNA binding strength; we hypothesize that this may be due to the relatively inert character of the Pt metal towards off-target biomolecules, facilitated by the steric hindrance created by the two axially oriented H atoms.

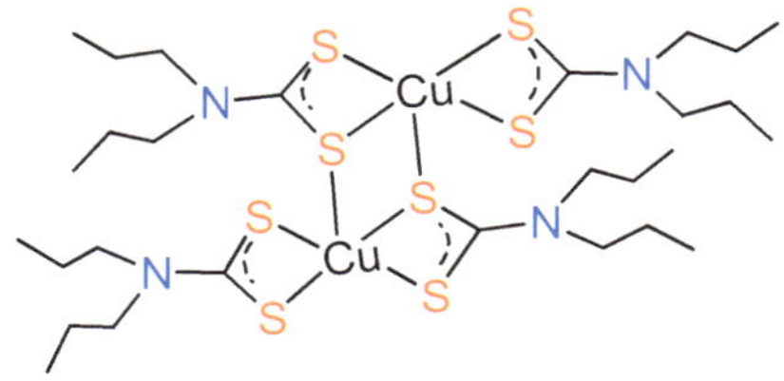

82

Fig. (10). Line drawing of Cu(II) dithiocarbamate complex **(82)**.

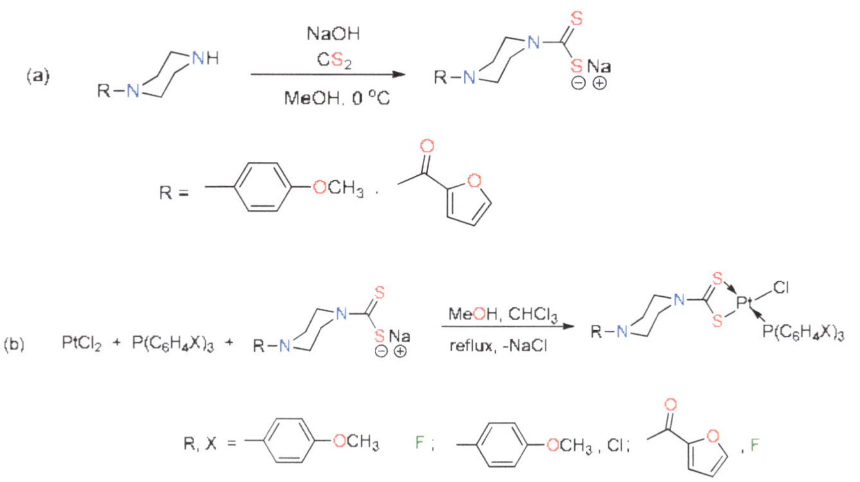

83 **84** **85**

Scheme 17. General synthetic route of piperazine-substituted dithiocarbamate ligands **(a)** and the Pt(II) **83-85** complexes **(b)**.

Rehman and his team reported two platinum(II) dithiocarbamate complexes, focusing on phenanthriplatin-type axial protection, equatorial-axial conformational isomerism, and their anticancer and DNA binding properties [54]. They synthesized platinum(II) dithiocarbamate complexes **86** and **87** are shown in Scheme **18**. Notably, complexes **86** and **87** displayed two distinct stereochemical conformations in the solid state, as determined by SCXRD, attributed to the flexible piperazine backbone. Although complex **87** has lower lipophilicity than complex **86**, it exhibits superior *in vitro* anticancer activity against selected cancer cell lines, emphasizing the importance of axial protection in enhancing anticancer efficacy over lipophilicity. Furthermore, the complexes show activity against the cancer-promoting protein NF-κB, suggesting that the mechanism of cancer cell death may involve the inhibition of NF-κB activity, leading to apoptosis. Previous studies using Annexin V-FITC have confirmed this apoptotic mode of cell death. Additionally, DNA-binding studies reveal that the complex-DNA adduct arrangement occurs spontaneously, with a non-intercalative mode of interaction driven by electrostatic and covalent interactions.

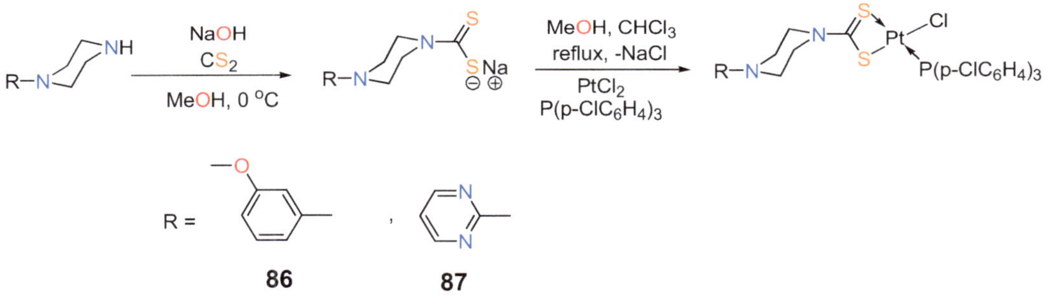

Scheme 18. General synthetic route of dithiocarbamate ligand and Pt(II) complexes **86-87**.

A summary of the key complexes, including their structures and IC_{50} values against various cell lines, is presented in Table **1**.

Table 1. A summary of the key complexes.

S. No.	Complexes	Structure Type	Tested Cell Line	Reported IC_{50}	Refs.
1.	Co(III) azepane-1-carbodithioate **(1)**	distorted octahedral (SCXRD)	MDA-MB-231 (Breast Cancer)		[30]
2.	Mn(II)/Co(II)/Cu(II)/Zn(II)/Pt(II) methoxyphenyl dithiocarbamato complexes **(5-9)**	bidentate ligand	MRC5-SV2 (human foetal lung cancer cell)	Mn: 31.5±3.8 µM, (SI 0.8); Co:12.9±3.8 µM (SI 1.4); Zn: 15.1±2.4µM(SI 2.0); Pt: 25.5±4.4 µM, (SI 0.2); Cisplatin: 8.6±0.3 µM, (SI 1.7)	[31]
3.	Ru(II) triphenylphosphine dithiocarbamate **(10-12)**	octahedral	breast (MCF7); renal (TK10; melanoma (UACC62)	IC_{50} (µM): 33.36 **(10)**, 38.18 **(11)**, 41.07 **(12)**; 79.39 **(10)**, 77.18 **(11)**, 91.95 **(12)**; 39.28 **(10)**, 72.32 **(11)**, 90.20 **(12)**	[32]

(Table 1) cont.....

S. No.	Complexes	Structure Type	Tested Cell Line	Reported IC_{50}	Refs.
4.	Sn(II) isophthaloylbis(piperazine-4,1-diyl) bis(methylene)dicarbamodithioate **(13-15)**	Six-coordinate geometry	anti-breast cancer/ anti-nematode activity	**14,15>13**	[33]
5.	Mn(II)/Cu(II)/Zn(II)/Pt(II) dithiocarbamate complexes **(16-19)**	distorted square-planar (Cu, Pt)	HeLa, MRC5(normal)	Cu/Zn=most efficient 30.1 ± 3.5/ 28.1 ± 1.8; Pt=least selective 39.7 ± 10.3	[34]
6.	Pd(II) and Pt(II) dithiocarbamate **(28,29)**	square-planar	MCF-7 (Breast Cancer)	Pd(II)=60μM; Pt~32μM	[40]
7.	Zn(II) dithiocarbamate **(33,35)**	distorted tetrahedral/ trigonal-bipyramidal/square-pyramidal	*Antileishmanial activity*	Significant anti-promastigote/amastigote IC_{50} (μM), 0.66 ± 0.22 **33**, 1.48 ± 0.10 **35**; 2.98 ± 0.30 **33**, 2.51 ± 0.10 **35**	[41]
8.	Bi(III) dithiocarbamate **(40,41)**	seven and eight coordinated Bi(III) center	*Antileishmanial activity*	Significant promastigote/amastigote IC_{50} (μM), 7.16/7.44, **40**; 8.40/9.70 **41**	[42]
9.	β-oxodithioester Pd(II) **48**,Pt(II) **50**	distorted square planar geometry	*Antileishmanial activity*	Significant anti-promastigote/anti-amastigote IC_{50} (μM), 0.59±0.10 **48**; 0.56±0.10 **50**; 0.85±0.272 **48**, 1.99± 0.08 **50**	[43]
10.	Pd(II) bis-dithiocarbamate (piperidinyl dithiocarbamate, phenylpiperazinyl, phenyl dithiocarbamate, benzyl piperadinyl) **(67-70)**	square-planar	HeLa, MCF-7, Caco-2	MCF-7 (μM): 8.09, 0.151, 0.11, and 0.032	[47]
11.	Cu(II) and Zn(II) morpholinyldithiocarbamato **(74-75)**	Cu = distorted square-planar, Zn = dimeric	TK-10, UACC62, MCF-7	Ligand (μM): 1.51, 12.73, 2.65 (respectively); **74** (μM): 4.64, 4.47, 4.37; **75** (μM): 8.70, 16.54, 3.17	[49]
12.	Cu(II) **76**, Zn(II) **77**, Pt(II) **78** bis(1-phenylpiperazine dithiocarbamate)	Zn=distorted tetrahedral, Pt=distorted square-planar	TK10; UACC6; MCF7	**76**: 19.83 3.34 17.52; **77**: >100 >100 >100; **78**: 13.40 15.14 8.42	[50]

CONCLUSION

This book chapter provides an in-depth overview of the versatile and promising class of metal dithiocarbamates, particularly those of Mn(II), Co(II)/(III), Ni(II), Cu(II), Zn(II), Pd(II), Pt(II), and Bi(III). Their facile synthesis and robust coordination chemistry enable the stabilization of diverse metal centers across various oxidation states. Numerous studies highlight their structural flexibility, with dithiocarbamate frameworks existing as monomeric, dimeric, or polymeric species, features that significantly influence their biological activity. These complexes exhibit potent cytotoxicity, Anti-Leishmanial Activity, DNA-cleavage ability, apoptosis induction, potential for drug delivery, effects on human breast cancer, anti-nematode activity, and antibacterial activity. Overall, this chapter emphasizes the rich chemical diversity and biomedical relevance of metal dithiocarbamates.

ACKNOWLEDGEMENTS

We sincerely acknowledge the financial support from the University of Allahabad (UoA) for providing the Seed Grant to K.K.M. We also express our gratitude to the Department of Chemistry, University of Allahabad, for the infrastructural facilities.

REFERENCES

[1] G. Hogarth, Transition metal dithiocarbamates: 1978-2003 progress in inorganic, chemistry 53 (2005) 71–561. [http://dx.doi.org/10.1002/0471725587]

[2] Karlin KD, Lippard SJ. Progress in Inorganic Chemistry. Wiley Online Library 2005.

[3] Sarker JC, Hogarth G. Dithiocarbamate complexes as single source precursors to nanoscale binary, ternary and quaternary metal sulfides. Chem Rev 2021; 121(10): 6057-123. [http://dx.doi.org/10.1021/acs.chemrev.0c01183] [PMID: 33847480]

[4] F. P. Andrew, P. A. Ajibade, Metal complexes of alkyl-aryl dithiocarbamates: Structural studies, anticancer potentials and applications as precursors for semiconductor nanocrystals, [http://dx.doi.org/10.1016/j.molstruc.2017.10.106]

[5] Raina-Fulton R. A review of methods for the analysis of orphan and difficult pesticides: glyphosate, glufosinate, quaternary ammonium and phenoxy acid herbicides, and dithiocarbamate and phthalimide fungicides. J AOAC Int 2014; 97(4): 965-77. [http://dx.doi.org/10.5740/jaoacint.SGERaina-Fulton] [PMID: 25145125]

[6] Berry DJ, Torres Martin de Rosales R, Charoenphun P, Blower PJ. Dithiocarbamate complexes as radiopharmaceuticals for medical imaging. Mini Rev Med Chem 2012; 12(12): 1174-83. [http://dx.doi.org/10.2174/138955712802762112] [PMID: 22931590]

[7] B.Y.A. El-Aarag, T. Kasai, M.A.H. Zahran, *et al. In vitro* anti-proliferative and anti-angiogenic activities of thalidomide dithiocarbamate analogs, Int. Immunopharmacol. 21 (2014) 283–292,

[8] Aghbash KO, Alamgholiloo H, Pesyan NN, Khaksar S, Rostamnia S. Gold nanoparticle stabilized dithiocarbamate functionalized magnetite carbon as promise clean nanocatalyst for A3-coupling organic transformation. Molecular Catalysis 2021; 499: 111252. [http://dx.doi.org/10.1016/j.mcat.2020.111252]

[9] Gao R, Li D, Zheng S, Gu H, Deng W. Colorimetric/fluorescent/Raman trimodal sensing of zinc ions with complexation-mediated Au nanorod. Talanta 2021; 225: 121975. [http://dx.doi.org/10.1016/j.talanta.2020.121975] [PMID: 33592723]

[10] Sathiyaraj E, Gurumoorthy G, Thirumaran S. Nickel(II) dithiocarbamate complexes containing the pyrrole moiety for sensing anions and synthesis of nickel sulfide and nickel oxide nanoparticles. New J Chem 2015; 39(7): 5336-49. [http://dx.doi.org/10.1039/C4NJ02250E]

[11] Hogarth G, Onwudiwe DC. Copper dithiocarbamates: coordination chemistry and applications in materials science. Inorganics (Basel) 2021; 9(9): 70. [http://dx.doi.org/10.3390/inorganics9090070]

[12] Ajiboye TO, Ajiboye TT, Marzouki R, Onwudiwe DC. The versatility in the applications of dithiocarbamates. Int J Mol Sci 2022; 23(3): 1317. [http://dx.doi.org/10.3390/ijms23031317] [PMID: 35163241]

[13] Lovejoy KS, Lippard SJ. Non-traditional platinum compounds for improved accumulation, oral bioavailability, and tumor targeting. Dalton Trans 2009; (48): 10651-9. [http://dx.doi.org/10.1039/b913896j] [PMID: 20023892]

[14] Kasherman Y, Sturup S, Gibson D. Is glutathione the major cellular target of cisplatin? A study of the interactions of cisplatin with cancer cell extracts. J Med Chem 2009; 52(14): 4319-28. [http://dx.doi.org/10.1021/jm900138u] [PMID: 19537717]

[15] Hogarth G. Metal-dithiocarbamate complexes: chemistry and biological activity. Mini Rev Med Chem 2012; 12(12): 1202-15. [http://dx.doi.org/10.2174/138955712802762095] [PMID: 22931592]

[16] Nabipour H, Ghammamy S, Ashuri S, Aghbolagh ZS. Synthesis of a new dithiocarbamate compound and study of its biological properties. Org Chem J 2010; 2: 75-80.

[17] Vojta D, Višnjevac A, Leka Z, Kosović M, Vazdar M. Temperature-induced release of crystal water in the Co, Mo and Pt complexes of N,N-diacetatedithiocarbamate. FTIR spectroscopy and quantum chemical study. J Mol Struct 2016; 1103: 245-53. [http://dx.doi.org/10.1016/j.molstruc.2015.09.038]

[18] P.J. Heard, Main group dithiocarbamate complexes. Progress Inorg. Chem. (2005) 53-69 2. [http://dx.doi.org/10.1002/0471725587.ch1]

[19] Sovilj SP, Avramović N, Katsaros N. Syntheses and properties of mixed dinuclear copper(II) complexes with heterocyclic dithiocarbamates and a cyclic octadentate tertiary amine. Trans Met Chem (Weinh) 2004; 29(7): 737-42. [http://dx.doi.org/10.1007/s11243-004-4995-6]

[20] Ajibade PA, Onwudiwe DC. Synthesis and characterization of group 12 complexes of *N, N* -methyl phenyl- *N, N* -butyl phenyl dithiocarbamate. J Coord Chem 2011; 64(17): 2963-73. [http://dx.doi.org/10.1080/00958972.2011.606906]

[21] Manar KK, Yadav MK, Anamika MGB, Drew MGB, Singh N. Influence of functionalities over polymer, trimer, dimer formation and optical properties of cadmium dithiocarbamates. Polyhedron 2016; 117: 592-9. [http://dx.doi.org/10.1016/j.poly.2016.06.047]

[22] Tamilvanan S, Gurumoorthy G, Thirumaran S, Ciattini S. Synthesis, characterization, cytotoxicity and antimicrobial studies on Bi(III) dithiocarbamate complexes containing furfuryl group and their use for the preparation of Bi2O3 nanoparticles. Polyhedron 2017; 121: 70-9. [http://dx.doi.org/10.1016/j.poly.2016.09.038]

[23] Hayat F, Rehman Z, Khan MH. Two new heteroleptic ruthenium(II) dithiocarbamates: synthesis, characterization, DFT calculation and DNA binding, J. Coord. Chem. 70 (2017) 279–295 10. S.H. Dar, S. Thirumaran, S. Selvanayagam, Synthesis, spectral and X-ray structural studies on Hg(II) dithiocarbamate complexes: A new precursor for HgS nanoparticles. Polyhedron 2015; 96: 16-24.

[24] Azizi N, Khajeh M, Hasani M, Dezfooli S. An efficient four-component synthesis of dithiocarbamate derivatives. Tetrahedron Lett 2013; 54(39): 5407-10. [http://dx.doi.org/10.1016/j.tetlet.2013.07.124]

[25] Li H, Lai CS, Wu J, Ho PC, de Vos D, Tiekink ERT. Cytotoxicity, qualitative structure–activity relationship (QSAR), and anti-tumor activity of bismuth dithiocarbamate complexes. J Inorg Biochem 2007; 101(5): 809-16. [http://dx.doi.org/10.1016/j.jinorgbio.2007.01.010] [PMID: 17363061]

[26] Tan YS, Ooi KK, Ang KP, *et al.* Molecular mechanisms of apoptosis and cell selectivity of zinc dithiocarbamates functionalized with hydroxyethyl substituents. J Inorg Biochem 2015; 150: 48-62. [http://dx.doi.org/10.1016/j.jinorgbio.2015.06.009] [PMID: 26086852]

[27] Desai KJ, Santhosh M, Akashvarunan P. Synthesis of S-aryl dithiocarbamates and its *in-silico* studies. Results Chem 2025; 15: 102257. [http://dx.doi.org/10.1016/j.rechem.2025.102257]

[28] Singh V, Chauhan R, Gupta AN, *et al.* Photosensitizing activity of ferrocenyl bearing Ni(ii) and Cu(ii) dithiocarbamates in dye sensitized TiO2 solar cells. Dalton Trans 2014; 43(12): 4752-61.

[http://dx.doi.org/10.1039/c3dt52142g] [PMID: 24473675]

[29] Onwudiwe DC, Arfin T, Strydom CA. Fe(II) and Fe(III) complexes of N-ethyl-N-phenyl dithiocarbamate: Electrical conductivity studies and Thermal properties. Electrochim Acta 2014; 127: 283-9.
[http://dx.doi.org/10.1016/j.electacta.2014.02.034]

[30] Jaiswal S, Rai N, Chandra S, *et al.* Design, synthesis, and structural evaluation of metal complexes of azepane-1-carbodithioate for targeting human breast cancer: investigating cytotoxic activity against the MDA-MB-231 cell line. New J Chem 2024; 48(30): 13589-601.
[http://dx.doi.org/10.1039/D4NJ01878H]

[31] Ajibade PA, Andrew FP, Fatokun AA, Oluwalana AE. Synthesis, characterization and *in vitro* screening for anticancer potential of Mn(II), Co(II), Cu(II), Zn(II), and Pt(II) methoxyphenyl dithiocarbamato complexes. J Mol Struct 2021; 1230: 129894.
[http://dx.doi.org/10.1016/j.molstruc.2021.129894]

[32] Ajibade PA, Andrew FP. Synthesis, characterization, and *in vitro* anticancer studies of chlorido(triphenylphosphine)ruthenium(II) dithiocarbamate complexes. Phosphorus Sulfur Silicon Relat Elem 2021; 196(9): 832-8.
[http://dx.doi.org/10.1080/10426507.2021.1925671]

[33] Kumar A, Parveen M, Ahamad S, *et al.* Synthesis & spectral studies of organotin(IV) dithiocarbamates derived from 2-aminoethyl piperazine: Anticancer & anti-nematode activity. J Mol Struct 2023; 1294: 136462.
[http://dx.doi.org/10.1016/j.molstruc.2023.136462]

[34] Ajibade PA, Fatokun AA, Andrew FP. Synthesis, characterization and anti-cancer studies of Mn(II), Cu(II), Zn(II) and Pt(II) dithiocarbamate complexes - crystal structures of the Cu(II) and Pt(II) complexes. Inorg Chim Acta 2020; 504: 119431.
[http://dx.doi.org/10.1016/j.ica.2020.119431]

[35] Singh A, Prasad LB, Shiv K, Kumar R, Garai S. Synthesis, characterization, and *in vitro* antibacterial and cytotoxic study of Co(II), Ni(II), Cu(II), and Zn(II) complexes of N-(4-methoxybenzyl) N-(phenylethyl) dithiocarbamate ligand. J Mol Struct 2023; 1288: 135835.
[http://dx.doi.org/10.1016/j.molstruc.2023.135835]

[36] Singh A, Kumar R, Shiv K, *et al.* Synthesis, crystal structure and screening for anticonvulsant and antianxiety activities of three new Ni(II), Cu(II), and Zn(II) dithiocarbamate complexes. J Mol Struct 2024; 1298: 137052.
[http://dx.doi.org/10.1016/j.molstruc.2023.137052]

[37] Eslami Moghadam M, Rezaeisadat M, Mansouri-Torshizi H, Hosseinzadeh S, Daneshyar H. New anticancer potential Pt complex with tertamyl dithiocarbamate ligand: Synthesis, DNA targeting behavior, molecular dynamic, and biological activity. J Mol Liq 2023; 379: 121651.
[http://dx.doi.org/10.1016/j.molliq.2023.121651]

[38] Hosseinzadeh S, Eslami Moghadam M, Sheshmani S, Shahvelayati AS. Some new anticancer platinum complexes of dithiocarbamate derivatives against human colorectal and pancreatic cell lines. J Biomol Struct Dyn 2020; 38(8): 2215-28.
[http://dx.doi.org/10.1080/07391102.2019.1627909] [PMID: 31204612]

[39] Hosseinzadeh S, Eslami Moghadam M, Sheshmani S, Shahvelayati AS. New anticancer Pd and Pt complexes of tertamyl dithiocarbamate and DACH ligands against HT29 and Panc1 cell lines. J Biomol Struct Dyn 2022; 40(15): 6910-20.
[http://dx.doi.org/10.1080/07391102.2021.1891972] [PMID: 33645454]

[40] Gokul G, Thirumaran S, Azhamuthu T. Synthesis, spectral, DFT and anticancer studies on bis(N-(3-methoxybenzyl)-N-(pyridin-3-ylmethyl)dithiocarbamato-S,S′)M(II) (M= Pd, Pt). J Indian Chem Soc 2024; 101(11): 101357.
[http://dx.doi.org/10.1016/j.jics.2024.101357]

[41] Manar KK, Yadav CL, Tiwari N, *et al.* Effect of functionalities on the crystal structures of new zinc(II) dithiocarbamates: a combined anti-leishmanial and thermal decomposition study. CrystEngComm 2017; 19(19): 2660-72.
[http://dx.doi.org/10.1039/C7CE00211D]

[42] Anamika A, Singh R, Manar KK, *et al.* Impact of substituents on the crystal structures and anti-leishmanial activity of new homoleptic Bi(III) dithiocarbamates. New J Chem 2019; 43(43): 16921-31.
[http://dx.doi.org/10.1039/C9NJ04477A]

[43] Yadav MK, Rajput G, Srivastava K, *et al.* Anti-leishmanial activity of Ni(II), Pd(II) and Pt(II) β-oxodithioester complexes. New J Chem 2015; 39(8): 6358-66.
[http://dx.doi.org/10.1039/C5NJ00765H]

[44] Yadav MK, Maurya AK, Rajput G, *et al.* New planar *trans* -copper(II) β-dithioester chelate complexes: synthesis, characterization, anticancer activity and DNA-binding/cleavage studies. J Coord Chem 2017; 70(4): 565-83.
[http://dx.doi.org/10.1080/00958972.2016.1275589]

[45] Yadav MK, Maurya AK, Rajput G, *et al.* Synthesis, characterization, DNA binding and cleavage activity of homoleptic zinc(II) β-oxodithioester chelate complexes. J Coord Chem 2017; 70(18): 3171-85.
[http://dx.doi.org/10.1080/00958972.2017.1377835]

[46] Yadav CL, Manar KK, Yadav MK, *et al.* Synthesis, crystal structures and properties of new homoleptic Ni(II)/Pd(II) β-oxodithioester chelates. J Mol Struct 2018; 1160: 488-96.
[http://dx.doi.org/10.1016/j.molstruc.2018.02.024]

[47] Paca AM, Ajibade PA, Andrew FP, Nundkumar N, Singh M. Synthesis, X-ray crystal structures and anticancer studies of four Pd(II) dithiocarbamate complexes. Arab J Chem 2021; 14(9): 103326.
[http://dx.doi.org/10.1016/j.arabjc.2021.103326]

[48] Singh RP, Singh A, Prasad LB, Shiv K, Hira SK, Manna PP. Nickel (II), Copper (II), and Zinc (II) Complexes of N-bis (4-methoxybenzyl) Dithiocarbamate: Synthesis, Characterization Studies, and Evaluation of Antitumor Activity. J Mol Struct 2022; 1264: 133295.
[http://dx.doi.org/10.1016/j.molstruc.2022.133295]

[49] Ajibade PA, Andrew FP, Botha NL, Solomane N. Synthesis, crystal structures and anticancer studies of morpholinyldithiocarbamato Cu(II) and Zn(II) complexes. Molecules 2020; 25(16): 3584.
[http://dx.doi.org/10.3390/molecules25163584] [PMID: 32781741]

[50] Andrew FP, Ajibade PA. Synthesis, characterization and anticancer studies of bis(1-phenylpiperazine dithiocarbamato) Cu(II), Zn(II) and Pt(II) complexes: Crystal structures of 1-phenylpiperazine dithiocarbamato-S,S′ zinc(II) and Pt(II). J Mol Struct 2018; 1170: 24-9.
[http://dx.doi.org/10.1016/j.molstruc.2018.05.068]

[51] F. P. Andrew and P. A. Ajibade, Synthesis, characterization and anticancer studies of bis-(N-methyl-1-phenyldithiocarbamato) Cu(II), Zn(II) and Pt(II) complexes. Single crystal X-ray structure of the copper complex, J. Mol. Coord., (2018).
[http://dx.doi.org/10.1080/00958972.2018.1489537]

[52] Ajibade PA, Mbuyazi TB, Paca AM, Andrew FP, Singh M. Morphological studies, photocatalytic degradation of organic dyes and anticancer potential of copper sulfide nanoparticles prepared from bis(diallyl dithiocarbamato)copper(II) complex. J Mol Struct 2024; 1312: 138537.
[http://dx.doi.org/10.1016/j.molstruc.2024.138537]

[53] Amir MK, Zia-ur-Rehman Z-R, Hayat F, *et al.* Monofunctional platinum(II) dithiocarbamate complexes: synthesis, characterization and anticancer activity. RSC Advances 2016; 6(112): 110517-24.
[http://dx.doi.org/10.1039/C6RA19469A]

[54] Imran M. Z. U. Rehman, G. Hogarth, *et al.* Two new monofunctional platinum(II) dithiocarbamate complexes: phenanthriplatin-type axial protection, equatorial-axial conformationan isomerism, and anticancer and DNA binding studies. Dalton Trans 2020; 49: 15385. [http://dx.doi.org/10.1039/D0DT03018J] [PMID: 33140800]

CHAPTER 4

Biological Potentials of Schiff Base Metal Complexes: A Multifaceted Approach

Mohan Kumar[1], **Madhuri Chaurasia**[2], **Deepak Tomar**[2,3] and **Sulekh Chandra**[4,*]

[1] *Department of Chemistry, M.M.H. College, Ghaziabad, Uttar Pradesh, India*

[2] *Department of Chemistry, University of Delhi, Delhi, India*

[3] *Department of Chemistry, RK(PG) College, Shamli, Uttar Pradesh, India*

[4] *Department of Chemistry, Zakir Husain Delhi College, University of Delhi, New Delhi, India*

Abstract: Metal-based drugs have been intensively explored in the pharmaceutical, biochemistry, agricultural, and clinical industries since the serendipitous discovery of the chemotherapeutic agent cisplatin. Coordination chemistry has been extensively studied because of its flexible bonding properties and diverse pharmacological properties. Metal complexes have exhibited a wide range of characteristics, including electrochemical, catalytic, photovoltaic, enzyme inhibitory, lipid-lowering properties, photodynamic therapy, and biological properties such as antibacterial, antitumor, antifungal, antioxidant, anticonvulsant, anti-inflammatory, antiviral, antifertility, anti-HIV, antiproliferative, and diuretic activities. Metal complexes have been shown to have higher biological activity than ligands. By introducing structural alterations to the ligands, the properties of the metal complexes will be improved. Metal complexes have distinctive characteristics, such as molecular geometries and ligand exchange, that enable them to interact with biomolecules in distinct ways and through different modes of action. Antibiotic resistance is a serious problem nowadays; thus, new drugs with a broad spectrum of action and no or minimal toxicity are needed. Metal complexes have low stability and solubility, making them difficult to introduce into clinical practice. Validation of metal-based drugs requires collaboration between chemists and biologists. In the present review, we focused on the recent development of metal complexes, including their synthesis, structure, and biological characteristics. We hope the present review will facilitate the development and introduction of more therapeutic drugs into clinical practice.

Keywords: Antimicrobial properties, Anticancer agents, Antibacterial, Antitumor, Antifungal, Antioxidant, Anticonvulsant, Bioinorganic chemistry, Coordination chemistry, Metal complexes, Schiff base.

* **Corresponding author Sulekh Chandra:** Department of Chemistry, Zakir Husain Delhi College, University of Delhi, New Delhi, India; E-mail: sc1953@gmail.com

Abhay Nanda Srivastva (Ed.)
All rights reserved-© 2026 Bentham Science Publishers

INTRODUCTION

Metals have A vital part in humans and may be found in many traces that are required for living things to exist. Since they are highly reactive and show a variety of coordination characteristics, transition metals can be employed to synthesize complexes, including a wide spectrum of organic ligands [1, 2]. The evolution of coordination chemistry has been significantly influenced by the characteristics and manufacture of such metal complexes. Over the past few decades, metal complexes have received much research about their possible application in the treatment of various deadly diseases [3 - 5]. Metal-based pharmaceuticals have emerged as one of the productive research fields in artificial biomedical sciences since the invention of the cisplatin pharmaceutical. In drug-derived metal complexes, through oxygen, nitrogen, and sulphur donor atoms in an ideal three-dimensional framework, the metal may coordinate ligands, allowing the molecule to detect and interact with a given biological target [6]. By introducing different groups to ligands, multiple chemical changes might increase this even further. Moreover, switching between different configurations is possible due to the enormous flexibility of coordination geometry, which is only determined by steric hindrance and ligand charge. Complexes of transition metals have demonstrated a substantial contribution to their use in the pharmaceutical sector. Metals are poisonous, non-biodegradable, and often accumulate in crucial human organs even at very high concentrations. The majority of these compounds exhibit a broad range of biologicalproperties, including antidiabetic [7, 8], anti-inflammatory [9], anticonvulsant [10], antidepressant [11], hypertensive [12], anthelmintic [13], antitubercular [14], anti-allergic [15], antifungal [16], antiviral [17], and antimicrobialproperties [18].

Schiff base complexes are furthermore well-known to show remarkable biological activity and serve as key models for physiologically significant entities, influenced by the presence of transition metal ions [19 - 21]. The -C=N-linkage in Schiff bases gives them biological and chemical significance. As a result of the nitrogen- and oxygen-giving atoms in Schiff bases, as well as their transition metal ions, the complexes they create have attracted a great deal of attention. A variety of biological effects have been reported for compounds with an aromatic ring containing one or more nitrogen atoms. It is estimated that over 75% of FDA-approved and currently available drugs contain heterocyclic nitrogen compounds [22, 23]. Numerous naturally occurring N-heterocyclic substances, which are widely dispersed, have physiological and pharmacological characteristics and are components of numerous biologically significant molecules, such as numerous antibiotics, vitamins, pharmaceuticals, dyes, agrochemicals, and nucleic acids, among many others, are among them [24].

There has been much interest in Schiff bases containing halogen groups and their metal complexes because of their antibacterial properties. Metal-based medications have grown rather important in the medical field recently and are now used to treat diseases, including diabetes, cancer, anti-inflammatory disorders, and cardiovascular diseases [25 - 27]. Many reports have been published on their application in homogeneous and heterogeneous catalysis in recent times. Many research studies have been conducted on the coordination behavior of these compounds because some metals have intrinsic chemical interests as multidentate ligands.

To give information on the present state of metal complexes and their biological profile, this review highlights literature that has been made public throughout the last few years. We hope that this study will provide a valuable opportunity to learn more about the metal complexes that are therapeutically relevant.

Biological Importance of Metal Complexes

The advancement of bioinorganic chemistry has raised interest in metal complexes, as numerous such complexes have been investigated as models for physiologically significant species. Biological molecules can bind to positively charged ions formed by metal complexes in aqueous solutions. Metal complexes may assemble into a myriad of coordination geometries, giving them distinct forms. Furthermore, a variety of unique molecular species that possess distinct coordination numbers and geometries might be provided by structurally modifying metal-based complexes. A metal complex or metal-based compound can coordinate with its ligand in a three-dimensional configuration, providing groups with the ability to be shaped to address defined molecular targets. They are distinctive for their synthetic versatility, selectivity, structure similarity, and azomethine group presence. Schiff bases are extensively studied. As a result, we've come to list a handful of their most crucial biological processes (Fig. **1**).

Antibacterial

The rise in the fatality rate of infectious illnesses is closely linked to certain bacteria that are resistant to antibiotics. The primary reason for this issue is the absence of workable solutions [28]. No doubt, an alarming situation arises for the creation of new antibacterial drugs, which have an effective mode of action. The bactericidal properties of Schiff bases are widely established. N-(salicylidene)-2-hydroxyaniline (Fig. **2**), for instance, has a MIC value of 8 I1/4mol/L and is effective against Mycobacterium TB H37Rv [29]. Tests on J774 macrophages were used to confirm compound 1's selectivity. Even at doses as high as 1000 I1/4mol/L, chemical 1 had no cytotoxic impact on J774 macrophages. Under these conditions, almost 80% of macrophages remained viable, demonstrating compound 1's excellent selectivity.

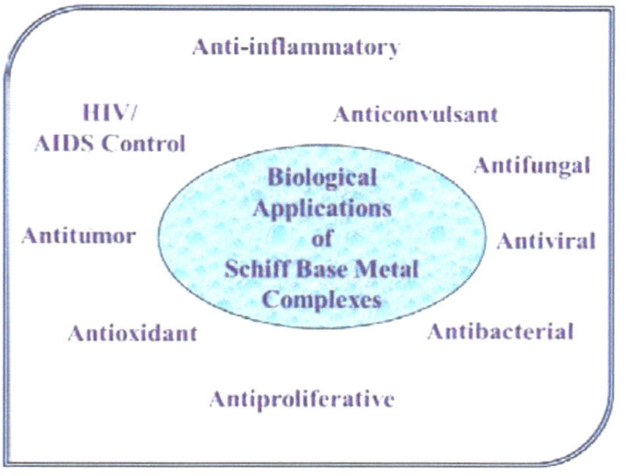

Fig. (1). Biological applications of schiff base metal complexes.

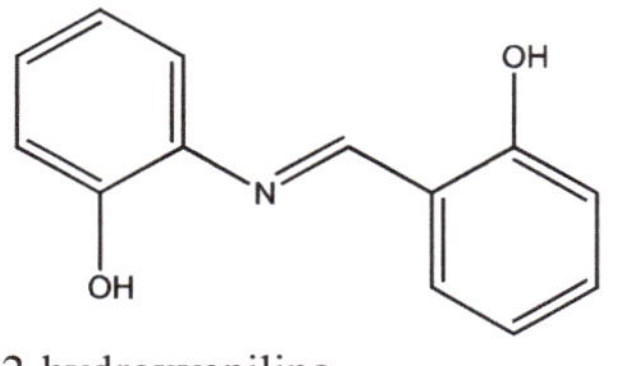

Fig. (2). Structure of N-(salicylidene)-2-hydroxyaniline.

Actinometry Madurahydroxylactones are secondary metabolites produced by the plant Rubra [30]. Examples of Schiff bases falling under this category include the imines 2-7 (Fig. **3**). Except for compounds 2 and 7, every product derived from madurahydroxylactone was effective in suppressing B *in vitro. Micrococcus flavus, S. aureus, S. subtilis*, and *Sarcina lutea* growth, with MIC values ranging from 0.2 to 3.1 I1/4mol/L.

With MIC values greater than 50.0 I1/4mol/L, all of these compounds had extremely limited action against Proteus vulgaris or Mycobacterium phlei [31]. Coumarins, amino acids, sulfonamides, aminothiazolylbromocoumarins, crown ethers or resacetophenones, O-phthaldehyde, 2-aminophenol, and 1,2,4-triazoles are additional compounds of natural or non-natural origin that serve as stand-ins for the synthesis of Schiff bases with antimicrobial properties [32 - 38]. The antibacterial characteristics of representative compounds from these classes were examined. However, they did not exhibit any significant activity.

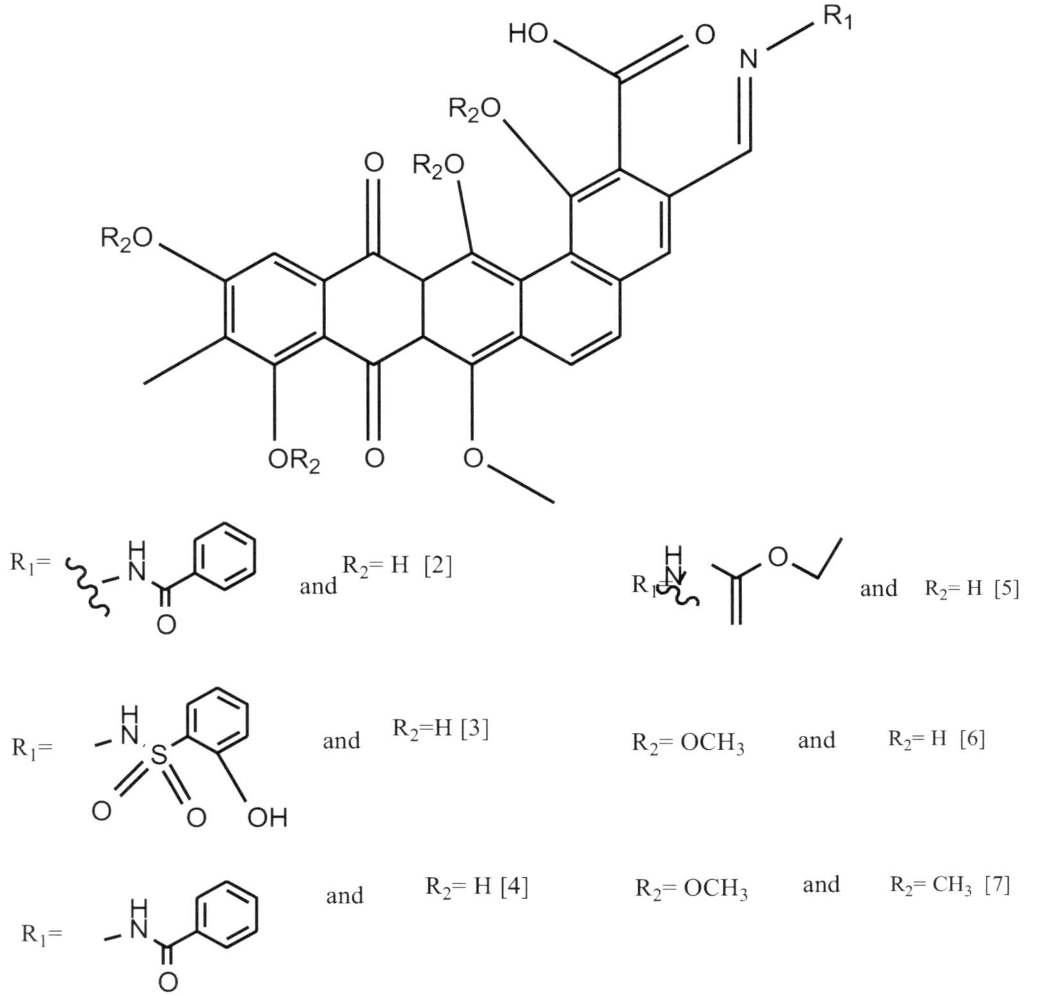

Fig. (3). Structure of Antibiotic Schiff bases derived from plant natural product.

Antitumor

The human body is prone to cancer due to several cells that replicate uncontrollably after a significant genetic mutation in healthy cells. Healthy cells can grow into cancerous cells as a result of mutations in genes that regulate cellular development, reproduction, and the cell cycle. These cells form larger populations, gain more bulk, skip cell checkpoints, avoid apoptosis, and begin attacking nearby, accessible organs. Cancer is a disease that continually affects people of all ages and also affects all organs and tissues. After affecting a certain organ, the illness spreads to other parts of the body [39].

Cancer can be effectively treated and controlled if found in an early stage. It can be deadly if not diagnosed and allowed to progress to an advanced stage. Chemotherapeutics are frequently rejected by more than 270 distinct malignancies, which leads to a recurrence of the disease after an initial cure. In order to solve this problem, new chemicals and techniques must be developed. Malignant cells have a diminished or halted apoptosis, a normal cell death mechanism for old, sick, and damaged cells that prevents the spread of cancer and can be restored by anticancer drugs. Therefore, a novel chemical's chances of successfully causing apoptosis in cancer cells during anticancer research are excellent [40].

Schiff bases, often referred to as azomethines, are produced when primary amines and aldehydes combine. These compounds have a range of biological, therapeutic, clinical, pharmacological, and analytical functions. Aromatic primary amines, which serve in regulating and encouraging biological processes, may contain additional donor functional groups like -Cl, -OH, $-CH_3$, *etc.*

Antifungal

Indeed, there has been an apparent increase in potentially fatal systemic fungal infections [41]. Fungal infections typically affect more than just the surface of tissues. The majority of the patients who are at risk are becoming more numerous, including those who have had major surgery, are on immunosuppressive medications, are getting cancer treatment, or have undergone solid-organ and hematopoietic stem cell transplants [42]. Strong antifungal medications should be developed and discovered, and numerous compounds of Schiff bases should be identified as having this potential [43, 44].

The phytopathogenic fungi Alternaria brassicae and Alternaria brassicicola harm the majority of cruciferous crops, such as cauliflower, broccoli, and N-(Salicylidene)-2-hydroxyaniline (Compound 8 in Fig. (**4**)) inhibited the growth of these fungi by 67-68% at 500 ppm [45]. Compounds 9 and 10 are examples of Schiff bases made from chitosan that exhibit antifungal properties (Fig. **4**). In the presence of 1000 ppm of these compounds, Botrytis cinerea and Colletotrichum lagenarium growth was inhibited by 26-33% and 35-38%, respectively. Only a few studies have examined the overall effect of Schiff bases on the growth of phytopathogenic fungi; these studies should be expanded. Schiff bases made from natural sources are promising for the development of brand-new antifungal drugs due to their antibacterial effect [46]. An intriguing method for creating a Schiff base generated from nystatin-dextran has been described by Domb *et al.* [47].

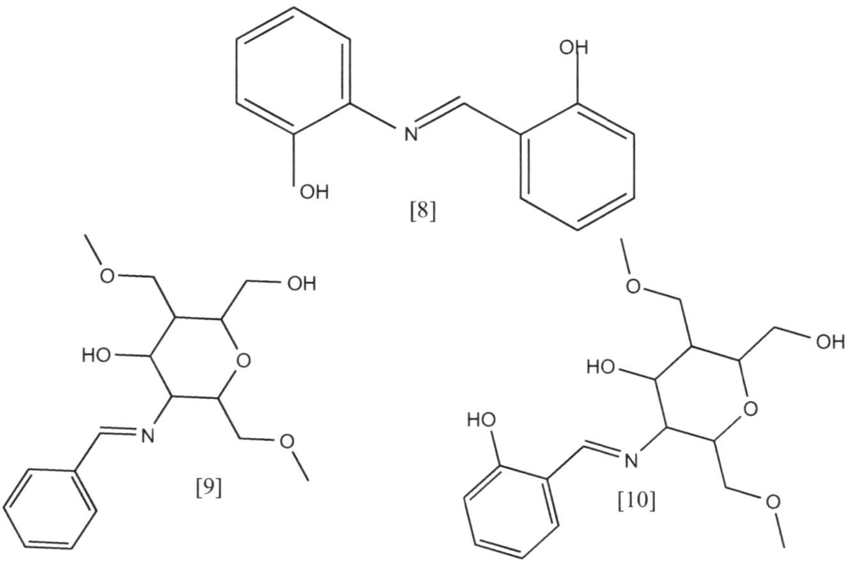

Fig. (4). Schiff base compounds having antifungal activities.

Antioxidant

There is a lot of interest in finding compounds with antioxidant properties. Antioxidants made synthetically are often utilized because they are more effective and less expensive than antioxidants made naturally. Numerous metal complexes of Schiff bases that function as antioxidants are capable of removing Reactive Oxygen Species (ROS) efficiently. Five chitosan and carboxymethyl chitosan Schiff bases' antioxidant capacities were examined using superoxide and hydroxyl radical scavenging devices.

Antioxidant compounds functioning as free radical scavengers, which play a vital role in the breakdown of food and chemical materials, may substantially delay or prevent the oxidation of easily oxidizable substrates. As a result, the need for finding antioxidants has greatly increased in recent years [48]. There has been a lot of research into these systems as a result of these results. The reaction of 3-hydroxysalicylidene-S-methylthiosemicarbazone with substitution R1 revealed the formation of complexes with nickel(II), iron(III), and oxovanadium(IV) (Fig. 5) [49]. An *in vitro* antioxidant evaluation of free ligands and metal complexes was carried out using the CUPRAC (CUPric Reducing Antioxidant Capability) approach.

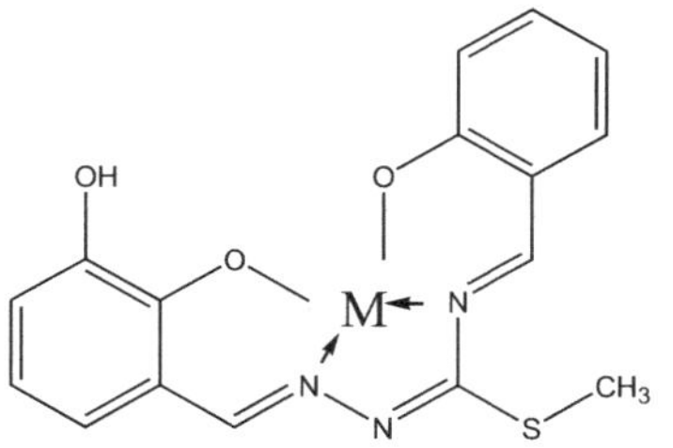

Fig. (5). 3-hydroxysalicylidene-S-methylthiosemicarbazone.

The approach is based on detecting the absorbance of Cu(I)-neocuproine, which is produced when the chain-breaking antioxidant undergoes a redox reaction with the CUPRAC reagent, Cu(II)-Nc [50]. In comparison to its complexes, the ligand demonstrated stronger *in vitro* antioxidant ability. The iron (III)complex has a greater Trolox equivalent antioxidant capacity rating than the other complexes. It was also examined whether free ligands and their complexes could scavenge reactive oxygen species like hydroxyl radicals, superoxide anion radicals, and hydrogen peroxide *in vitro*. This revealed that VO and Fe complexes in particular had noteworthy scavenging action for reactive oxygen species [51].

Compared to benzene Schiff bases, the antioxidant capacity of o-(1-ferrocenyl-ethylidene amino) phenol, m-(1-ferrocenylethylidene amino), and p-(1-ferrocenyl-ethylidene amino) phenol was significantly boosted by the addition of the ferrocenyl group [52]. Bovine Serum Albumin (BSA) has been found to bind Cu, Ni, Co, and Zn complexes of glutamic acid-salicylaldehyde, indicating that this binding boosted BSA's antioxidant activity by more than ten times [53]. Through the formation of many hydrogen and van der Waals bonds, a molecular docking investigation revealed that Schiff bases were bound with BSA, encircled by residues of tryptophan [54]. Binuclear copper(II) complexes of Schiff bases and HAS were studied in relation to one another [55]. Researchers looked into how Bovine Serum Albumin (BSA) and cobalt(II) and zinc(II) complexes of N-2-hydroxyacetophenon-N0-2-hydroxynaphthaldehyde-1,2-phenylenediimine inter-acted [56].

Several keto-amine Schiff bases produced from 8-hydroxyquinoline were assessed as *in vitro* antioxidants [57]. By using the phosphomolybdenum technique, a number of substituted-N-[(1E)-phenylmethylidene]benzohydrazide analogues were created and assessed for their *in vitro* antioxidant properties [58]. The antioxidant capabilities of the Schiff base N-(2-hydroxyl acetophenone)- 3-oxapentane-1,5-diamine and its Ni-complex were assessed [59]. Researchers have looked into the antioxidant properties of cobalt(II) and copper(II) complexes produced through the condensation of polyamine with 2-hydroxybenzaldehyde or 2-hydroxy-3-methoxybenzaldehyde [60].

Research has been conducted on the antioxidant property of copper complex synthesis from 2-((4-(2-amino benzyl)-1,4-diazepam-1-yl)methyl)benzenamine and 2-formylpyridine [61]. More powerful antioxidant activity was seen in Fe(II), Co(II), Ni(II), Ru(II), and Pd(II) complexes synthesized from 2-hydroxy-3-methoxybenzaldehyde and methyl 6-acetamide-2-amino-4,5-6,7-tetrahydrobenzo[b]thiophene-3-carboxylate [62, 63]. By reacting 1-hydroxy-2-acetonapthanone with pentane-1,3-diamine and propane-1,3-diamine, Schiff bases were created, and they showed potential antioxidant properties [64]. Few titanium(IV) complexes of di-negative tridentate Schiff base ligands (ONO, aroylhydrazine with aldehyde, or ketone) have been studied for their antioxidant and microbiological activities [65]. Dioxomolybdenum(VI) complexes were synthesized by reacting 2-hydroxybenzophenone, 2-hydroxy-4-methoxybenzophenone, 2-hydroxy-4-octyloxybenzophenone, 2-hydroxy-4-methoxy-40-methylbenzophenone, and 2-hydroxy-4-allyloxy benzophenone with thiocarbohydrazide [66]. These compounds show excellent antioxidant capacities for the detoxification of superoxide radicals, and the Cu(II) complex of N-(2-hydroxy-1-naphthaldehyde)-glucosamine has been proposed [67].

Anticonvulsant

Epidemiological studies estimate that more than sixty million people worldwide suffer from epilepsy, a condition marked by recurring seizures [68]. The usual pattern of brain activity is disturbed in epilepsy, leading to aberrant feelings, behaviours, and perceptions. Occasionally, there may also be convulsions, muscle spasms, and unconsciousness. The brain condition that causes the aforementioned ailments is one in which groups of neurons or nerve cells in the brain occasionally signal improperly. A large number of antiepileptic drugs have been in practice since 1985. They frequently have noticeable negative side effects and do not completely control all individuals' seizures [69, 70]. For medicinal chemistry researchers, then, looking for safer and stronger antiepileptic medications is both important and difficult. A vital part of epilepsy treatment regimens, antiepileptic drugs (AEDs) mostly aim to control seizures without producing side effects. In the realm of medical inorganic chemistry, there is now a discernible surge in interest in the synthesis of metal compounds as medications and diagnostic instruments [71]. The speciation of metal species in biological media according to their potential interactions with biomolecules is a primary focus of this study in order to provide novel therapies and diagnostics [72]. 4-amino Numerous metal complexes have been discovered to have a wide variety of applications in the fields of biology, medicine, analysis, and pharmacology, including antipyrine and its complexes. These chemicals include radiosensitizing agents, antibacterials, antivirals, antiparasitics, metal-mediated antibiotics, and anticancer medications [73]. Biochemists are presently interested in research on a novel class of chemotherapeutic chemicals. In experimental epilepsy models, such as the Maximum Electroshock test (MES) and subcutaneous pentylenetetrazol (sc-PTZ)-induced seizures in mice, the produced compounds were assessed as anticonvulsant drugs [74]. They were also tested for neurotoxicity using a mouse rotarod assay.

A class of Schiff bases of phthalimide, 4-(1,3-dioxo-1,3-dihydro-2H-isoindol-2-yl)-N-(substitutedphenyl)methylene/ethyli-dene benzohydrazide, was investigated for its neurotoxic and anticonvulsant properties [75]. The findings indicated that every chemical tested was both more active and less harmful than the widely used medication phenytoin. Researchers have found that anticonvulsants containing nitro groups at the ortho position of distal aryl rings are the most potent. Lamotrigine was synthesized using isatin and substituted isatin in a series of 3,3 [6-(2,3-dichlorophenyl)-1,2,4-triazine-3,5-diyl]dinitrolobis-(substituted-1,3-dihy-dro-2Hindol-2-one), and their anticonvulsant action was tested [76]. A maximum electroshock seizure test was conducted using lamotrigine and phenobarbitone sodium as standard medications. The produced compounds have a stronger anticonvulsant effect than regular lamotrigine, according to the results [77]. There have been several brand-new 3-aryl-4(3H)-quinazolinones-2-carboxaldehyde as well as their related Schiff bases and thiosemicarbazone derivatives created. Since the thiosemicarbazone side chain terminates with a fluorine atom and a free amino group, molecules may have cytotoxic, analgesic, and anticonvulsant properties [78].

Anti-inflammatory

When examined for anti-inflammatory efficacy, the dipropylenetriamine Schiff mono-base with 2-thiophene-carboxaldehyde decreased the oedema that carrageenin induced in rat paws (52.0-82.6%) [79]. The most effective anti-inflammatory drug, N-(4-bromobenzylidene)-[2-(2,6-dichloro aniline)benzyl carbazide], was free of gastrointestinal toxicity. The anti-inflammatory, analgesic, and antipyretic properties of 4-aminophenazone (5-dimethyl-2-phenyl-1,2-dihydro-3H-pyrazol-3-one) Schiff base derivatives were investigated in mice. The paw oedema was inhibited by all test drugs in a dose-dependent manner [80]. Schiff bases made from thiazoles exhibit analgesic and anti-inflammatory properties [81, 82]. During their synthesis, -4-(3-substituted-4-sydnonylidene) amino-5-mercapto-1,2,4-triazoles' analgesic and anti-inflammatory qualities were examined. Amino-5-mercapto-1,2,4-triazole's analgesic and anti-inflammatory properties were enhanced by the electron-releasing group in sydnone [83]. The most effective anti-inflammatory substance was a Schiff base made from 2,3-dihydroxybenzaldehyde and 4-amino-1,5-dimethyl-2-phenylpyrazol-3-one [84]. Making a series of S-substituted phenacyl 1,3,4-oxadiazoles and Schiff bases from 2-[(2,6-dichloroanilino) phenyl] acetic acid (diclofenac acid) and screening them biologically. Due to their diverse biological actions, the 1,3,4-oxadiazoles and diclofenac moieties are considered to be of significant importance [85].

Antiviral

Though biological sciences have advanced recently, one of the most significant worldwide health concerns still is viral infection [86]. Viral illnesses are widespread and can range from mild to fatal infections, as seen by the recent COVID-19 epidemic. In contrast to bacterial and fungal infection disorders, a general appropriate measure capable of controlling viral infections is still lacking in treatment [87]. The Fe (III) and Ni (II) complexes containing thiosemicarbazones were synthesized, and their inhibitory activities against SARS-CoV-23CL-like protease were examined using experimental and theoretical approaches [88]. The iron complex was more effective than the nickel complex, demonstrating that it can be a powerful inhibitor of such enzymes. The synthesis of complexes of Co^{3+} with substituted biguanide derivatives as ligands and their antiviral properties against herpes simplex virus type II has been reported [89]. The number of free functional groups present in ligands seems to influence the antiviral activity more than the metal complex structures do. The antiviral activities of two novel iridium (III) complexes containing substituted phenyl pyridine or phenylquinoline and a chelating ligand were investigated [90]. The two compounds have also exhibited antiviral potency against Herpes simplex viruses HSV-1 and HSV-2, where interaction with the major groove of the HSV double helix was seen. Using Non-Steroidal Anti-Inflammatory Medications (NSAIDs) as ligands, Elena and her colleagues produced Cu (II) complexes and studied their potential as an antiviral agent against SARS-CoV-2 target proteins [91]. According to the study, these metal complexes may be able to bind to SARS-CoV-2 target proteins that are essential for the development of COVID-19. The target proteins include RNA-dependent RNA-polymerase of COVID-19, papain-like protease, Nsp16-Nsp10 2'-O-methyltransferase complex, and 3C-like cysteine major protease. An extensive literature review on potent platinum-based metal complexes has been performed by Federica *et al.* [92]. The extensive experimental results presented may raise new issues and perspectives, indicating that platinum compounds should be considered not only as anticancer drugs, but also as antiviral agents. Researchers have synthesized a metal complex containing coumarin and have shown that it inhibits both COVID-19 and the SARS virus [93]. An additional exploratory approach is required to fully understand the biological mechanisms of antiviral action, investigate the potential clinical effects of the compounds, and determine the potential therapeutic applications of these compounds against diverse viruses (Fig. **6**).

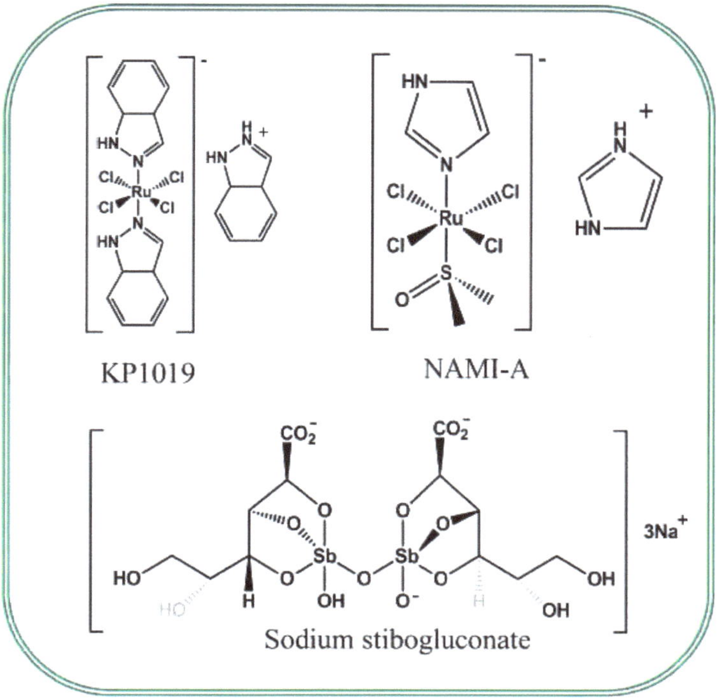

Fig. (6). A Few important metal-based antiviral drugs.

HIV/AIDS Control

Today, there is a critical challenge with Acquired Immunodeficiency Syndrome (AIDS), one of the most chronic ailments caused by the human immunodeficiency virus (HIV) [94, 95]. According to estimates from the World According to the World Health Organization (WHO), there are around 2 million new cases of HIV detected each year, and about 38 million people are living with the virus globally [96, 97]. The growth of viral mutations and their severe side effects creates major challenges in the treatment of HIV infection; therefore, the an urgent requirement to develop creative anti-HIV drugs which are vital for HIV/AIDS treatment. As anti-HIV medications, terpyridyl Ru (II) complexes have been extensively studied [98]. The more active Ru (II) complex (RuTz2) may selectively seek out TAR RNA by chemical recognition of H-bonds, further stabilizing the Ru (II)-RNA complex through electrostatic attraction. As HIV-1 reverse transcriptase inhibitors, mononuclear Cu (II) heterocyclic Schiff base complexes have been synthesized and their activity has been evaluated using the HIV-1 RT kit assay [99]. A new class of polypyridyl Ru (II) complexes has been developed, which inhibit both Moloney murine leukemia virus (M-MuLV) and HIV-1 reverse transcriptase effectively [100]. Due to molecular recognition of hydrogen bonds, these complexes can firmly and selectively bind HIV-1 TAR RNA, further stabilizing the system that is bound to Ru(II). Cationic meso-thiophenium porphyrins and their zinc complexes have been synthesized without photodynamic reactions [101]. It was found that $T_2(OH)_2M$ (A_2B_2-type) and $T(OH)_3MZn$ (AB_3-type) Zn(II)-complexes were able to inhibit HIV-1 entry to a significant extent, with $T_2(OH)_2MZn$ showing the most effective anti-HIV activity. Under non-PDT circumstances, it was proposed that the anti-HIV and antibacterial properties of the A2B2-type porphyrins were significantly enhanced by the addition of two thiophenium groups at their meso-positions in conjunction with zinc. Xu *et al.* carried out an in-depth review and explored the potential of coumarin derivatives as anti-HIV therapeutics [102]. To enable the logical design of more successful anti-HIV candidates, they also discuss the Structure-Activity Relationship (SAR) and mechanisms of action very well. A recent review has been published where authors discussed more than 7 metal complexes in detail, including gold, platinum, ruthenium, zinc, *etc.* [103]. With various degrees of activity and efficacy, platinum and ruthenium compounds have lately been studied for their potential use as antimycobacterial and anti-HIV medicines and are nicely reviewed by Yufani and their co-workers [104].

Antiproliferative

It is a therapeutic method for slowing or stopping the growth of cells, including cancerous cells, in nearby tissues. Wongsuwan and colleagues employed the A549 human lung adenocarcinoma cell line to investigate the anticancer activity of Schiff base ligands (N-(8-quinolyl)-X-salicylaldimine) and their complexes with iron (II) and iron (III) [105]. The Fe (III) complex had better anticancer activity against the A549 cells (IC50 = 10 μM) than the two well-known anticancer medications, etoposide and cisplatin (IC50 = 19 and 16 AμM, respectively). Alyar *et al.* described the synthesis and biological assessment of new Schiff bases that include sulfisoxazole, sulfamethoxazole, and their copper (II) complexes. The sulfonamide B test was used to screen for anticancer activity against three human cancer cell lines. Promising cytotoxic effects against breast (MCF7) cells were demonstrated by Cu(S2M-S1)2 and Cu(S1M-S1)2 complexes with S1M-S1 (IC50 = 40 AμM *vs.* 5-fluorouracil, 5-FU, IC50 = 0.10 AμM) [106]. In addition to this, docking investigations revealed that hydrogen bonds are critical for reversible interactions between proteins and small molecules. Hydrogen bonds are expected to form when nitrogen and oxygen atoms are connected to the benzene ring, the O-atom on the sulfisoxazole ring, and the O-atom on the sulfonyl groups. Schiff base ligands made with Co(II) and Ru(II) and their M(II) complexes were investigated by researchers [107]. The compounds' anticancer properties were evaluated *in vitro* against the human colon cancer cell line (Caco-2) in addition to examining their biocompatibility in the normal fibroblast cell line (L-929). It was demonstrated that both complexes had a mild cytotoxic impact on L-929 cells but had anticancer efficacy against Caco-2 cells (IC50 = 610 AμM for the Co complex). Maria Ruiz and her colleagues investigated the antiproliferative properties of many copper (II) phenanthroline complexes in a study using colorectal cancer cell types [108]. The antiproliferative effects of copper complexes (MMP) can be attributed to the generation of ROS (reactive oxygen species) and the modification of mitochondrial membrane potential. *In vitro*, antiproliferation experiments conducted by Lopez and his collaborators demonstrated that bismuth(III) complexes showed a strong antiproliferative effect on breast cancer cell types [109]. Researchers have synthesized novel heterobimetallic nickel(II) ferrocenyldithiophosphonato complexes and examined their antiproliferative abilities against hepatitis, colon, and breast cancer cells [110]. MTT assays were used to evaluate the cytotoxic activity of each complex on three cancerous cell lines (HepG2, DLD-1, and MDA-MB-231). It was found that all nickel complexes were cytotoxic to the MDA-MB-231 cell line. The complexes $[Ni(FcL1)_2]$ and $[Ni(FcL2)_2]$ are cytotoxic to DLD-1 and HepG2 cells, but compounds $[Ni(FcL3)_2]$ and $[Ni(FcL4)_2]$ are not. Therefore, substituents on molecules' structures have a significant influence on their activity. Ni (II), Pd (II), and Zn (II) thiosemicarbazone complexes were evaluated by Carcelli and colleagues for their antiproliferative properties on human histiocytic lymphoma U937 cells [111]. It was claimed that the structural changes made to the thiosemicarbazone skeleton have no impact on the antiproliferative effect when comparing the four nickel compounds' cytotoxic efficacy. Rudbari *et al.* developed asymmetric halogen-substituted [NNO] ligand complexes and screened their activity on the cancer cells A2780 and HCT116 [112]. In A2780 cells, I_2-H_2L_2-Cu complex induced cell death mechanisms, including apoptosis and autophagy, due to its antiproliferative effect. There was a direct relation between high cytotoxicity and intracellular ROS production. I_2-H_2L_2-Cu complex has been found to trigger ROS extrinsically, as indicated by the increase in intracellular ROS without destabilizing mitochondrial membranes, supporting an extrinsic apoptotic response. We have shown a few recently reported antiproliferative metal complexes in Fig. (7).

CONCLUSION

Here, we've demonstrated that the chemical composition and reactivities of metal-based medications vary drastically and are exclusively dependent on the characteristics of the metal core and the ligands around it. We highly advise that metal-based pharmaceuticals be included as much as feasible in new drug development testing protocols due to the huge chemical variation that makes metal complexes, at least in theory, a rich repository of innovative medicinally valuable molecules. Due to well-founded worries about their systemic toxicity, this medication group has seen a drop in attention and clinical use over the past few decades. It is worth noting, however, that each metal center has unique properties that can be used to successfully treat a specific pharmacological target as well as a particular disease. Inorganic drugs have therefore remained in clinical use for decades and persist because they have unique pharmacological properties that organic drugs cannot mimic. A more traditional method of finding novel drugs for metal-based compounds might be used, in which collections of these compounds are first examined *in vitro* and then *in vivo* in suitable experimental animals with the illness. In contrast to organic molecules, metal complexes have unique modes of action because of their complex molecular morphologies, propensity for ligand exchange reactions, and accessibility to redox activities. Eventually, additional research will likely result in metal-based inhibitors that have enhanced affinity and selectivity, providing an alternative to the currently explored orthodox therapies.

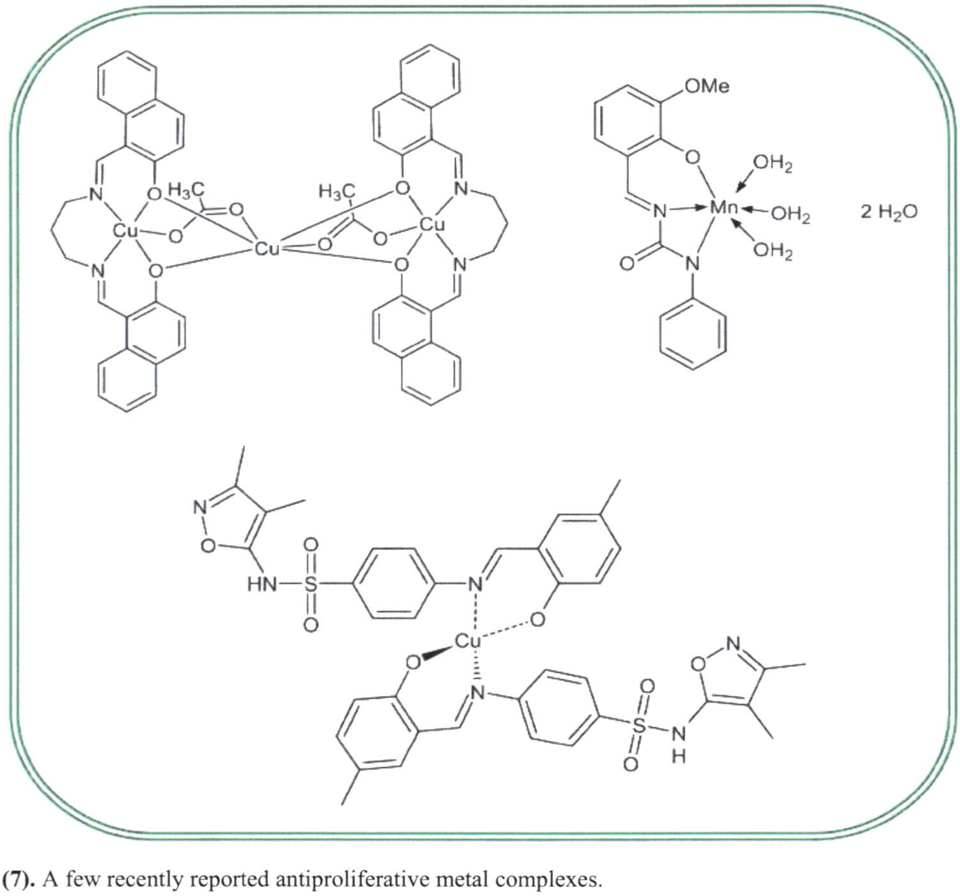

Fig. (7). A few recently reported antiproliferative metal complexes.

ACKNOWLEDGEMENTS

The authors gratefully acknowledge M.M.H. College, Ghaziabad, affiliated to Chaudhary Charan Singh University, Meerut, U.P., and Rashtriya Kisan Post Graduate College, affiliated to Maa Shakambhari University, Saharanpur, for providing a research platform.

REFERENCES

[1] El-Sonbati AZ, El-Mogazy MA, Nozha SG, *et al.* Mixed ligand transition metal(II) complexes: Characterization, spectral, electrochemical studies, molecular docking and bacteriological application. J Mol Struct 2022; 1248: 131498.
[http://dx.doi.org/10.1016/j.molstruc.2021.131498]

[2] Chhikara A, Tomar D, Bartwal G, *et al.* Thiadiazole functionalized salicylaldehyde-schiff base as a pH-responsive and chemo-reversible "turn-off" fluorescent probe for selective Cu (II) detection: Logic gate behaviour and molecular docking studies. Journal of Fluorescence 2023; 33(1): 25-41.

[3] Marichev KO, Patil SA, Patil SA, Heras Martinez HM, Bugarin A. *N* -heterocyclic carbene metal complexes as therapeutic agents: a patent review. Expert Opin Ther Pat 2022; 32(1): 47-61.
[http://dx.doi.org/10.1080/13543776.2021.1965992] [PMID: 34365884]

[4] Hamza RZ, Sheshah ZA, Suleman RH, Al-Juaid NF, Hamed NA, Al-Juaid MA. Efficacy of some antibiotics and some metal complexes (Nano-formula) that could increase their effectiveness during COVID-19. International Journal of Biological and Pharmaceutical Sciences Archive 2022; 3(1): 008-14.
[http://dx.doi.org/10.53771/ijbpsa.2022.3.1.0021]

[5] Kumar, M., Tomar, D., Chaurasia, M., Kumari, A., & Chandra, S. Synthesis, spectroscopic characterization, DFT-D2 analysis, and BSA binding studies of Co (II), Ni (II), and Cu (II) complexes with a novel Benzothiazole-based ligand. Transit. Met. Chem., 2025, 1-19.

[6] Bessen NP, Jackson JA, Jensen MP, Shafer JC. Sulfur donating extractants for the separation of trivalent actinides and lanthanides. Coord Chem Rev 2020; 421: 213446.
[http://dx.doi.org/10.1016/j.ccr.2020.213446]

[7] Kaushal R, Kaur M. Bio-medical potential of chalcone derivatives and their metal complexes as antidiabetic agents: a review. J Coord Chem 2021; 74(4-6): 725-42.
[http://dx.doi.org/10.1080/00958972.2021.1875450]

[8] Deswal Y, Asija S, Dubey A, *et al.* Cobalt(II), nickel(II), copper(II) and zinc(II) complexes of thiadiazole based Schiff base ligands: Synthesis, structural characterization, DFT, antidiabetic and molecular docking studies. J Mol Struct 2022; 1253: 132266.
[http://dx.doi.org/10.1016/j.molstruc.2021.132266]

[9] Saritha TJ, Metilda P. Synthesis, spectroscopic characterization and biological applications of some novel Schiff base transition metal (II) complexes derived from curcumin moiety. J Saudi Chem Soc 2021; 25(6): 101245.
[http://dx.doi.org/10.1016/j.jscs.2021.101245]

[10] Najm S, Naureen H, Sultana K, *et al.* Schiff-based metal complexes of lamotrigine: Design, synthesis, characterization, and biological evaluation. ACS Omega 2021; 6(11): 7719-30.
[http://dx.doi.org/10.1021/acsomega.1c00027] [PMID: 33778282]

[11] Yuldasheva, N., Acikyildiz, N., Akyuz, M., Yabo-Dambagi, L., Aydin, T., Cakir, A., & Kazaz, C. The Synthesis of Schiff bases and new secondary amine derivatives of p-vanillin and evaluation of their neuroprotective, antidiabetic, antidepressant, and antioxidant potentials. J. Mol. Str., 2022, 133883.

[12] Winiarski JP, de Barros MR, Wecker GS, *et al.* A novel hybrid organic-inorganic silsesquioxane and

cobalt(II) tetrasulphophthalocyanine material as an efficient electrochemical sensor for the simultaneous determination of the anti-hypertensive nifedipine and its metabolite. J Mater Chem C Mater Opt Electron Devices 2020; 8(20): 6839-50.
[http://dx.doi.org/10.1039/D0TC00429D]

[13] Patil SA, Kandathil V, Sobha A, *et al.* Comprehensive review on medicinal applications of coumarin-derived imine-metal complexes. Molecules 2022; 27(16): 5220.
[http://dx.doi.org/10.3390/molecules27165220] [PMID: 36014460]

[14] Mikwa CC, Toh-Boyo GM, Njong RN, *et al.* Bivalent metal complexes of a novel modified nicotinic acid hydrazide drug: Synthesis, characterization, and anti-tubercular studies. Eur J Chem 2022; 13(1): 63-8.
[http://dx.doi.org/10.5155/eurjchem.13.1.63-68.2183]

[15] Katouah HA, Al-Fahemi JH, Elghalban MG, *et al.* Synthesis of new Cu(II)-benzohydrazide nanometer complexes, spectral, modeling, CT-DNA binding with potential anti-inflammatory and anti-allergic theoretical features. Mater Sci Eng C 2019; 96: 740-756, 740-756.
[http://dx.doi.org/10.1016/j.msec.2018.11.034] [PMID: 30606587]

[16] Jaafar A, Mansour N, Fix-Tailler A, *et al.* Synthesis, characterization, antibacterial and antifungal activities evaluation of metal complexes with benzaldehyde-4-methylthiosemicarbazone derivatives. ChemistrySelect 2022; 7(10): e202104497.
[http://dx.doi.org/10.1002/slct.202104497]

[17] Shamim S, Gul S, Rauf A, *et al.* Gemifloxacin-transition metal complexes as therapeutic candidates: antimicrobial, antifungal, anti-enzymatic, and docking studies of newly synthesized complexes. Heliyon 2022; 8(8): e10378.
[http://dx.doi.org/10.1016/j.heliyon.2022.e10378] [PMID: 36061017]

[18] Deswal Y, Asija S, Kumar D, *et al.* Transition metal complexes of triazole-based bioactive ligands: synthesis, spectral characterization, antimicrobial, anticancer and molecular docking studies. Res Chem Intermed 2022; 48(2): 703-29.
[http://dx.doi.org/10.1007/s11164-021-04621-5]

[19] Keshavarzian E, Asadi Z, Poupon M, Dusek M, Rastegari B. Heterodinuclear Cu-Gd (3d-4f) complex with di-compartmental Schiff base ligand in biological activity: Synthesis, crystal structure, catecholase activity and DNA & BSA-binding studies. J Mol Liq 2022; 345: 117785.
[http://dx.doi.org/10.1016/j.molliq.2021.117785]

[20] El-Sonbati AZ, Diab MA, El-Sayed AK, Abou-Dobara MI, El-Sayad SA. Synthesis, characterization, molecular docking, biological activity, and DNA cleavage studies of Cu(II), Co(II), Ni(II), Mn(II), and Cd(II) Schiff base complexes. Appl Organomet Chem 2022; 36(5): e6623.
[http://dx.doi.org/10.1002/aoc.6623]

[21] Jasińska A, Szklarzewicz J, Jurowska A, *et al.* V(III) and V(IV) Schiff base complexes as potential insulin-mimetic compounds - Comparison, characterization and biological activity. Polyhedron 2022; 215: 115682.
[http://dx.doi.org/10.1016/j.poly.2022.115682]

[22] Tiwary BK, Pradhan K. Heterocyclic scaffolds in novel synthetic antibacterial agents. Alternatives to Antibiotics. Singapore: Springer 2022; pp. 223-42.
[http://dx.doi.org/10.1007/978-981-19-1854-4_9]

[23] Kumari S, Maddeboina K, Bachu RD, Boddu SHS, Trippier PC, Tiwari AK. Pivotal role of nitrogen heterocycles in Alzheimer's disease drug discovery. Drug Discov Today 2022; 27(10): 103322.
[http://dx.doi.org/10.1016/j.drudis.2022.07.007] [PMID: 35868626]

[24] Srivastava V, Singh PK, Tivari S, Singh PP. Visible light photocatalysis in the synthesis of pharmaceutically relevant heterocyclic scaffolds. Org Chem Front 2022; 9(5): 1485-507.
[http://dx.doi.org/10.1039/D1QO01602D]

[25] Sen S, Won M, Levine MS, *et al.* Metal-based anticancer agents as immunogenic cell death inducers:

the past, present, and future. Chem Soc Rev 2022; 51(4): 1212-33.
[http://dx.doi.org/10.1039/D1CS00417D] [PMID: 35099487]

[26] Li R, Hou X, Li L, Guo J, Jiang W, Shang W. Application of metal-based nanozymes in inflammatory disease: A review. Front Bioeng Biotechnol 2022; 10: 920213.
[http://dx.doi.org/10.3389/fbioe.2022.920213] [PMID: 35782497]

[27] Scarim CB, de Farias RL, Chiba DE, Chin CM. Insight into recent drug discoveries against trypanosomatids and plasmodium spp parasites: New metal-based compounds. Curr Med Chem 2022; 29(13): 2334-81.
[http://dx.doi.org/10.2174/0929867328666210917114912] [PMID: 34533436]

[28] da Silva CM, da Silva DL, Modolo LV, *et al.* Schiff bases: A short review of their antimicrobial activities. J Adv Res 2011; 2(1): 1-8.
[http://dx.doi.org/10.1016/j.jare.2010.05.004]

[29] Souza A O D, Galetti F, Silva C L, Bicalho B, Parma M M, Fonseca S F, *et al.* Quim Nova 2007; 30: 1563-6.

[30] Ding T, Yang LJ, Zhang WD, Shen YH. The secondary metabolites of rare actinomycetes: chemistry and bioactivity. RSC Advances 2019; 9(38): 21964-88.
[http://dx.doi.org/10.1039/C9RA03579F] [PMID: 35518871]

[31] Kim BM, Choi HY, Kim GW, Zheng CJ, Kim YH, Kim WG. Madurahydroxylactone, an inhibitor of *Staphylococcus aureus* FtsZ from Nonomuraea sp. AN100570. J Microbiol Biotechnol 2017; 27(11): 1994-8.
[http://dx.doi.org/10.4014/jmb.1708.08044] [PMID: 28910864]

[32] Chohan ZH, Arif M, Sarfraz M. Metal-based antibacterial and antifungal amino acid derived Schiff bases: their synthesis, characterization and *in vitro* biological activity. Appl Organomet Chem 2007; 21(4): 294-302.
[http://dx.doi.org/10.1002/aoc.1200]

[33] Baluja S, Solanki A, Kachhadia N. Evaluation of biological activities of some Schiff bases and metal complexes. J Indian Chem Soc 2006; 3(4): 312-7.
[http://dx.doi.org/10.1007/BF03245952]

[34] Venugopala KN, Jayashree BS. Microwave-induced synthesis of schiff bases of aminothiazolyl bromocoumarins as antibacterials. Indian J Pharm Sci 2008; 70(1): 88-91.
[http://dx.doi.org/10.4103/0250-474X.40338] [PMID: 20390087]

[35] Yildiz M, KÍraz A, DA1/4lger B. Synthesis and antimicrobial activity of new crown ethers of Schiff base type. J Serb Chem Soc 2007; 72(3): 215-24.
[http://dx.doi.org/10.2298/JSC0703215Y]

[36] Abdallah SM, Mohamed GG, Zayed MA, El-Ela MSA. Spectroscopic study of molecular structures of novel Schiff base derived from o-phthaldehyde and 2-aminophenol and its coordination compounds together with their biological activity. Spectrochim Acta A Mol Biomol Spectrosc 2009; 73(5): 833-40.
[http://dx.doi.org/10.1016/j.saa.2009.04.005] [PMID: 19505840]

[37] T'ang A, Lien EJ, Lai MMC. Optimization of the Schiff bases of N-hydroxy-N'-aminoguanidine as anticancer and antiviral agents. J Med Chem 1985; 28(8): 1103-6.
[http://dx.doi.org/10.1021/jm00146a022] [PMID: 2991520]

[38] Bayrak H, Demirbas A, Karaoglu SA, Demirbas N. Synthesis of some new 1,2,4-triazoles, their Mannich and Schiff bases and evaluation of their antimicrobial activities. Eur J Med Chem 2009; 44(3): 1057-66.
[http://dx.doi.org/10.1016/j.ejmech.2008.06.019] [PMID: 18676062]

[39] Uddin N, Rashid F, Ali S, *et al.* Synthesis, characterization, and anticancer activity of Schiff bases. J Biomol Struct Dyn 2020; 38(11): 3246-59.

[http://dx.doi.org/10.1080/07391102.2019.1654924] [PMID: 31411114]

[40] Crowley LC, Scott AP, Marfell BJ, Boughaba JA, Chojnowski G, & Waterhouse NJ. Measuring cell death by propidium iodide uptake and flow cytometry. Cold Spring Harb. Protoc., 2016, 2016(7), pdb-prot087163.
[http://dx.doi.org/10.1101/pdb.prot087163]

[41] Sundriyal S, Sharma R, Jain R. Current advances in antifungal targets and drug development. Curr Med Chem 2006; 13(11): 1321-35.
[http://dx.doi.org/10.2174/092986706776873023] [PMID: 16712473]

[42] Nucci M, Marr KA. Emerging fungal diseases. Clin Infect Dis 2005; 41(4): 521-6.
[http://dx.doi.org/10.1086/432060] [PMID: 16028162]

[43] Martins CVB, da Silva DL, Neres ATM, *et al.* Curcumin as a promising antifungal of clinical interest. J Antimicrob Chemother 2008; 63(2): 337-9.
[http://dx.doi.org/10.1093/jac/dkn488] [PMID: 19038979]

[44] Martins CVB, de Resende MA, da Silva DL, *et al.* *In vitro* studies of anticandidal activity of goniothalamin enantiomers. J Appl Microbiol 2009; 107(4): 1279-86.
[http://dx.doi.org/10.1111/j.1365-2672.2009.04307.x] [PMID: 19486403]

[45] Rehman W, Baloch MK, Muhammad B, Badshah A, Khan KM. Characteristic spectral studies and *in vitro* antifungal activity of some Schiff bases and their organotin (?) complexes. Chin Sci Bull 2004; 49(2): 119-22.
[http://dx.doi.org/10.1360/03wb0174]

[46] Ahmad A, Abid O U R, Rehman W, *et al.* J Chem Chem Eng Research Article Vol 2020; 39(2).

[47] Domb AJ, Linden G, Polacheck I, Benita S. Nystatin-dextran conjugates: Synthesis and characterization. J Polym Sci A Polym Chem 1996; 34(7): 1229-36.
[http://dx.doi.org/10.1002/(SICI)1099-0518(199605)34:7<1229::AID-POLA9>3.0.CO;2-Y]

[48] Domingo JL. Vanadium and tungsten derivatives as antidiabetic agents: a review of their toxic effects. Biol Trace Elem Res 2002; 88(2): 097-112.
[http://dx.doi.org/10.1385/BTER:88:2:097] [PMID: 12296430]

[49] Pisoschi AM, Cheregi MC, Danet AF. Total antioxidant capacity of some commercial fruit juices: electrochemical and spectrophotometrical approaches. Molecules 2009; 14(1): 480-93.
[http://dx.doi.org/10.3390/molecules14010480] [PMID: 19158657]

[50] Torabi S, Mohammadi M, Shirvani M. Antidiabetic, antioxidant, antibacterial, and antifungal activities of vanadyl Schiff base complexes. Trends Pharmacol Sci 2018; 4(2).

[51] Abdel Aziz AA, Elbadawy HA. Spectral, electrochemical, thermal, DNA binding ability, antioxidant and antibacterial studies of novel Ru(III) Schiff base complexes. Spectrochim Acta A Mol Biomol Spectrosc 2014; 124: 404-15.
[http://dx.doi.org/10.1016/j.saa.2014.01.050] [PMID: 24508879]

[52] Li YF, Liu ZQ. Ferrocenyl Schiff base as novel antioxidant to protect DNA against the oxidation damage. Eur J Pharm Sci 2011; 44(1-2): 158-63.
[http://dx.doi.org/10.1016/j.ejps.2011.07.001] [PMID: 21782019]

[53] Wang RM, Mao JJ, Song JF, Huo CX, He YF. Antioxidant activity of bovine serum albumin binding amino acid Schiff-bases metal complexes. Chin Chem Lett 2007; 18(11): 1416-8.
[http://dx.doi.org/10.1016/j.cclet.2007.09.015]

[54] Dezhampanah H, Firouzi R, Moradi Shoeili Z, Binazir R. Intermolecular investigation on interaction of two ternary copper(II) Schiff base complexes with bovine serum albumin. J Mol Struct 2020; 1205: 127557.
[http://dx.doi.org/10.1016/j.molstruc.2019.127557]

[55] Jiang S, Ni H, Liu F, Gu S, Yu P, Gou Y. Binuclear Schiff base copper(II) complexes: Syntheses,

crystal structures, HSA interaction and anti-cancer properties. Inorg Chim Acta 2020; 499: 119186. [http://dx.doi.org/10.1016/j.ica.2019.119186]

[56] Sedighipoor M, Kianfar AH, Sabzalian MR, Abyar F. Synthesis and characterization of new unsymmetrical Schiff base Zn (II) and Co (II) complexes and study of their interactions with bovin serum albumin and DNA by spectroscopic techniques. Spectrochim Acta A Mol Biomol Spectrosc 2018; 198: 38-50. [http://dx.doi.org/10.1016/j.saa.2018.02.050] [PMID: 29505940]

[57] Sashidhara KV, Kumar A, Bhatia G, Khan MM, Khanna AK, Saxena JK. Antidyslipidemic and antioxidative activities of 8-hydroxyquinoline derived novel keto-enamine Schiffs bases. Eur J Med Chem 2009; 44(4): 1813-8. [http://dx.doi.org/10.1016/j.ejmech.2008.08.004] [PMID: 18819730]

[58] Bala S, Uppal G, Kamboj S, Saini V, Prasad DN. Design, characterization, computational studies, and pharmacological evaluation of substituted-N'-[(1E) substituted-phenylmethylidene]benzohydrazide analogs. Med Chem Res 2013; 22(6): 2755-67. [http://dx.doi.org/10.1007/s00044-012-0270-0]

[59] Wu H, Jia F, Kou F, Liu B, Yuan J, Bai Y. A Schiff base ligand N-(2-hydroxylacetophenone--3-oxapentane-1,5-diamine and its nickel(II) complex: synthesis, crystal structure, antioxidation, and DNA-binding properties. Trans Met Chem (Weinh) 2011; 36(8): 847-53. [http://dx.doi.org/10.1007/s11243-011-9539-2]

[60] Soroceanu A, Bargan A. Advanced and biomedical applications of Schiff-base ligands and their metal complexes: A review. Crystals (Basel) 2022; 12(10): 1436. [http://dx.doi.org/10.3390/cryst12101436]

[61] Keypour H, Aidi M, Mahmoudabadi M, Karamian R, Asadbegy M, Gable RW. Synthesis, X-ray crystal structural, antioxidant and antibacterial studies of new Cu(II) macroacyclic Schiff base complex with a ligand containing homopiperazine moiety. J Mol Struct 2019; 1198: 126666. [http://dx.doi.org/10.1016/j.molstruc.2019.06.024]

[62] BingAl M, Turan N. Schiff base and metal(II) complexes containing thiophene-3-carboxylate: Synthesis, characterization and antioxidant activities. J Mol Struct 2020; 1205: 127542. [http://dx.doi.org/10.1016/j.molstruc.2019.127542]

[63] Yeğiner G, GA1/4lcan M, Işık S, Ürüt GÖ, Özdemir S, Kurtoğlu M. Transition metal (II) complexes with a novel azo-azomethine Schiff base ligand: Synthesis, structural and spectroscopic characterization, thermal properties and biological applications. J Fluoresc 2017; 27(6): 2239-51. [http://dx.doi.org/10.1007/s10895-017-2166-3] [PMID: 28840399]

[64] Vhanale BT, Deshmukh NJ, Shinde AT. Synthesis, characterization, spectroscopic studies and biological evaluation of Schiff bases derived from 1-hydroxy-2-acetonapthanone. Heliyon 2019; 5(11): e02774. [http://dx.doi.org/10.1016/j.heliyon.2019.e02774] [PMID: 31763472]

[65] Uddin MN, Khandaker S, Moniruzzaman , *et al.* Synthesis, characterization, molecular modeling, antioxidant and microbial properties of some Titanium(IV) complexes of schiff bases. J Mol Struct 2018; 1166: 79-90. [http://dx.doi.org/10.1016/j.molstruc.2018.04.025]

[66] Kaya Y, Erçağ A, Kaya K. Synthesis, characterization and antioxidant activities of dioxomolybdenum(VI) complexes of new Schiff bases derived from substituted benzophenones. J Coord Chem 2018; 71(20): 3364-80. [http://dx.doi.org/10.1080/00958972.2018.1516872]

[67] Mir JM, Kumar Rajak D, Charitra Maurya R. Bio-conjugated N-(2-hydroxy-1-naphthaldeh-de)-glucosamine Cu(II) complex: Bacterial sensitivity and superoxide dismutase-like activity. J Coord Chem 2018; 71(14): 2225-42. [http://dx.doi.org/10.1080/00958972.2018.1482488]

[68] Zhang T, Liu Z, Sun X, *et al.* Botany, traditional uses, phytochemistry, pharmacological and toxicological effects of *Croton tiglium* Linn.: a comprehensive review. J Pharm Pharmacol 2022; 74(8): 1061-84.
[http://dx.doi.org/10.1093/jpp/rgac040] [PMID: 35723937]

[69] Perucca E. An introduction to antiepileptic drugs. Epilepsia 2005; 46(s4) (Suppl. 4): 31-7.
[http://dx.doi.org/10.1111/j.1528-1167.2005.463007.x] [PMID: 15968807]

[70] Al-Soud YA, Al-Masoudi NA, Ferwanah AERS. Synthesis and properties of new substituted 1,2,4-triazoles: potential antitumor agents. Bioorg Med Chem 2003; 11(8): 1701-8.
[http://dx.doi.org/10.1016/S0968-0896(03)00043-9] [PMID: 12659756]

[71] Thompson KH, Orvig C. Metal complexes in medicinal chemistry: new vistas and challenges in drug design. Dalton Trans 2006; (6): 761-4.
[http://dx.doi.org/10.1039/B513476E] [PMID: 16437168]

[72] Timerbaev AR, Hartinger CG, Aleksenko SS, Keppler BK. Interactions of antitumor metallodrugs with serum proteins: advances in characterization using modern analytical methodology. Chem Rev 2006; 106(6): 2224-48.
[http://dx.doi.org/10.1021/cr040704h] [PMID: 16771448]

[73] Komiya N, Murahashi SI. Transition Metal-Catalyzed C-H Oxidation of Saturated Hydrocarbons with Molecular Oxygen. Chem Rec 2021; 21(8): 1928-40.
[http://dx.doi.org/10.1002/tcr.202100154] [PMID: 34235852]

[74] Sakthivel A, Jeyasubramanian K, Thangagiri B, Raja JD. Recent advances in schiff base metal complexes derived from 4-aminoantipyrine derivatives and their potential applications. J Mol Struct 2020; 1222: 128885.
[http://dx.doi.org/10.1016/j.molstruc.2020.128885]

[75] Bhat MA, Al-Omar MA. Synthesis, characterization and *in vivo* anticonvulsant and neurotoxicity screening of Schiff bases of phthalimide. Acta Pol Pharm 2011; 68(3): 375-80.
[PMID: 21648191]

[76] Kajal A, Bala S, Kamboj S, Sharma N, & Saini V. Schiff bases: A versatile pharmacophore. J. Catal., 2013, 2013.

[77] Kulkarni AA, Wankhede SB, Dhawale ND, Yadav PB, Deore VV, Gonjari ID. Synthesis, characterization and biological behavior of some Schiff's and Mannich base derivatives of Lamotrigine. Arab J Chem 2017; 10: S184-9.
[http://dx.doi.org/10.1016/j.arabjc.2012.07.020]

[78] Aly MM, Mohamed YA, El-Bayouki KAM, Basyouni WM, Abbas SY. Synthesis of some new 4(3H)-quinazolinone-2-carboxaldehyde thiosemicarbazones and their metal complexes and a study on their anticonvulsant, analgesic, cytotoxic and antimicrobial activities - Part-1. Eur J Med Chem 2010; 45(8): 3365-73.
[http://dx.doi.org/10.1016/j.ejmech.2010.04.020] [PMID: 20510483]

[79] Pontiki E, Hadjipavlou-Litina D, Chaviara AT. Evaluation of anti-inflammatory and antioxidant activities of Copper (II) Schiff mono-base and Copper(II) Schiff base coordination compounds of dien with heterocyclic aldehydes and 2-amino-5-methyl-thiazole. J Enzyme Inhib Med Chem 2008; 23(6): 1011-7.
[http://dx.doi.org/10.1080/14756360701841251] [PMID: 19005947]

[80] Uddin MN, Ahmed SS, Alam SMR. REVIEW: Biomedical applications of Schiff base metal complexes. J Coord Chem 2020; 73(23): 3109-49.
[http://dx.doi.org/10.1080/00958972.2020.1854745]

[81] Bhosale D, Narale A, Hadimani P, *et al.* Novel Schiff base derivatives containing 4,5-disubstituted thiazole as potential antibiofilm, anti-inflammatory and antioxidant agents: Green synthesis, molecular docking and ADME analysis. J Mol Struct 2024; 1311: 138401.

[http://dx.doi.org/10.1016/j.molstruc.2024.138401]

[82] Guo Z, Xing R, Liu S, *et al.* The synthesis and antioxidant activity of the Schiff bases of chitosan and carboxymethyl chitosan. Bioorg Med Chem Lett 2005; 15(20): 4600-3. [http://dx.doi.org/10.1016/j.bmcl.2005.06.095] [PMID: 16122924]

[83] Nithinchandra , Kalluraya B, Aamir S, Shabaraya AR. Regioselective reaction: Synthesis, characterization and pharmacological activity of some new Mannich and Schiff bases containing sydnone. Eur J Med Chem 2012; 54: 597-604. [http://dx.doi.org/10.1016/j.ejmech.2012.06.011] [PMID: 22795833]

[84] Alam MS, Choi JH, Lee DU. Synthesis of novel Schiff base analogues of 4-amino-1,5-dimeth-l-2-phenylpyrazol-3-one and their evaluation for antioxidant and anti-inflammatory activity. Bioorg Med Chem 2012; 20(13): 4103-8. [http://dx.doi.org/10.1016/j.bmc.2012.04.058] [PMID: 22626550]

[85] Bhandari SV, Bothara KG, Raut MK, Patil AA, Sarkate AP, Mokale VJ. Design, synthesis and evaluation of antiinflammatory, analgesic and ulcerogenicity studies of novel S-substituted phenacyl-1,3,4-oxadiazole-2-thiol and Schiff bases of diclofenac acid as nonulcerogenic derivatives. Bioorg Med Chem 2008; 16(4): 1822-31. [http://dx.doi.org/10.1016/j.bmc.2007.11.014] [PMID: 18248993]

[86] LeA3n-Figueroa DA, Bonilla-Aldana DK, Pachar M, *et al.* The never-ending global emergence of viral zoonoses after COVID-19? The rising concern of monkeypox in Europe, North America and beyond. Travel Med Infect Dis 2022; 49: 102362. [http://dx.doi.org/10.1016/j.tmaid.2022.102362] [PMID: 35643256]

[87] Bo W, Ahmad Z, Alanzi ARA, Al-Omari AI, Hafez EH, Abdelwahab SF. The current COVID-19 pandemic in China: An overview and corona data analysis. Alex Eng J 2022; 61(2): 1369-81. [http://dx.doi.org/10.1016/j.aej.2021.06.025]

[88] Atasever Arslan B, Kaya B, Şahin O, Baday S, Saylan CC, AolkA1/4seven B. The iron(III) and nickel(II) complexes with tetradentate thiosemicarbazones. Synthesis, experimental, theoretical characterization, and antiviral effect against SARS-CoV-2. J Mol Struct 2021; 1246: 131166. [http://dx.doi.org/10.1016/j.molstruc.2021.131166] [PMID: 34316082]

[89] Kirin VP, Demkin AG, Sukhikh TS, Ilyicheva TN, Maksakov VA. Cobalt complexes with biguanide derivatives - Synthesis, structure and antiviral activity. J Mol Struct 2022; 1250: 131486. [http://dx.doi.org/10.1016/j.molstruc.2021.131486]

[90] Bai D, Tian Y, Chen K, *et al.* Novel 6F-Iridium(III) complexes as potent theranostic agents: Hypoxia probing, radiosensitization and antiviral functionalities. Dyes Pigments 2020; 182: 108635. [http://dx.doi.org/10.1016/j.dyepig.2020.108635]

[91] Geromichalou EG, Trafalis DT, Dalezis P, Malis G, Psomas G, Geromichalos GD. *In silico* study of potential antiviral activity of copper(II) complexes with non-steroidal anti-inflammatory drugs on various SARS-CoV-2 target proteins. J Inorg Biochem 2022; 231: 111805. [http://dx.doi.org/10.1016/j.jinorgbio.2022.111805] [PMID: 35334392]

[92] De Castro F, De Luca E, Benedetti M, Fanizzi FP. Platinum compounds as potential antiviral agents. Coord Chem Rev 2022; 451: 214276. [http://dx.doi.org/10.1016/j.ccr.2021.214276]

[93] Ashraf A, El-Desouky MG, El-Afify MA. Thermal and spectroscopic studies of some prepared metal complexes and investigation of their potential anticancer and antiviral drug activity against SARS-CoV-2 by molecular docking simulation. Biointerface Res Appl Chem 2021; 12(1): 1053-75. [http://dx.doi.org/10.33263/BRIAC121.10531075]

[94] Uwishema O, Ayoub G, Badri R, *et al.* Neurological disorders in HIV: Hope despite challenges. Immun Inflamm Dis 2022; 10(3): e591. [http://dx.doi.org/10.1002/iid3.591] [PMID: 35146953]

[95] Barin F. HIV / AIDS as a model for emerging infectious disease: Origin, dating and circumstances of an emblematic epidemiological success. Presse Med 2022; 51(3): 104128. [http://dx.doi.org/10.1016/j.lpm.2022.104128] [PMID: 35623545]

[96] Khatri G, Sahito AM, Siddiqui MS, *et al.* HIV's imminent menace in a COVID-ridden country: The case of Pakistan. Ann Med Surg (Lond) 2022; 78: 103869. [http://dx.doi.org/10.1016/j.amsu.2022.103869] [PMID: 35734670]

[97] Liblik K, Byun J, Saldarriaga C, Mendoza I, & Baranchuk A. HIV & Heart. In neglected tropical diseases and other infectious diseases affecting the heart. AP, 2022, 125-136. [http://dx.doi.org/10.1016/B978-0-323-91122-1.00012-X]

[98] Guo YX, Liu M, Zhou YQ, Bi XD, Gao F. Terpyridyl ruthenium complexes as visible spectral probe for poly(A) RNA and bifunctional TAR RNA binders and HIV-1 reverse transcriptase inhibitors. Inorg Chim Acta 2022; 539: 121027. [http://dx.doi.org/10.1016/j.ica.2022.121027]

[99] Shanty AA, Raghu KG, Mohanan PV. Synthesis, characterization: Spectral and theoretical, molecular docking and *in vitro* studies of copper complexes with HIV RT enzyme. J Mol Struct 2019; 1197: 154-63. [http://dx.doi.org/10.1016/j.molstruc.2019.06.097]

[100] Wang MF, Li Y, Bi XD, *et al.* Polypyridyl ruthenium complexes as bifunctional TAR RNA binders and HIV-1 reverse transcriptase inhibitors. J Inorg Biochem 2022; 234: 111880. [http://dx.doi.org/10.1016/j.jinorgbio.2022.111880] [PMID: 35679745]

[101] Sengupta D, Rai M, Hoque Mazumdar Z, *et al.* Two cationic meso-thiophenium porphyrins and their zinc-complexes as anti-HIV-1 and antibacterial agents under non-photodynamic therapy (PDT) conditions. Bioorg Med Chem Lett 2022; 65: 128699. [http://dx.doi.org/10.1016/j.bmcl.2022.128699] [PMID: 35341921]

[102] Xu Z, Chen Q, Zhang Y, Liang C. Coumarin-based derivatives with potential anti-HIV activity. Fitoterapia 2021; 150: 104863. [http://dx.doi.org/10.1016/j.fitote.2021.104863] [PMID: 33582266]

[103] Nareetsile F, Matshwele JT, Ndlovu S, Ngaski M. Transition metal complexes with HIV/AIDS inhibitory properties. Chem Rev Lett 2020; 3(3): 140-60.

[104] Yufanyi DM, Abbo HS, Titinchi SJJ, Neville T. Platinum(II) and Ruthenium(II) complexes in medicine: Antimycobacterial and Anti-HIV activities. Coord Chem Rev 2020; 414: 213285. [http://dx.doi.org/10.1016/j.ccr.2020.213285]

[105] Wongsuwan S, Chatwichien J, Pinchaipat B, *et al.* Synthesis, characterization and anticancer activity of Fe(II) and Fe(III) complexes containing N-(8-quinolyl)salicylaldimine Schiff base ligands. J Biol Inorg Chem 2021; 26(2-3): 327-39. [http://dx.doi.org/10.1007/s00775-021-01857-9] [PMID: 33606116]

[106] Alyar S, A-zmen AoA-, Adem Ş, Alyar H, Bilen E, Kaya K. Synthesis, spectroscopic characterizations, carbonic anhydrase II inhibitory activity, anticancer activity and docking studies of new Schiff bases of sulfa drugs. J Mol Struct 2021; 1223: 128911. [http://dx.doi.org/10.1016/j.molstruc.2020.128911]

[107] Alkış ME, Keleştemür Ü, Alan Y, Turan N, Buldurun K. Cobalt and ruthenium complexes with pyrimidine based schiff base: Synthesis, characterization, anticancer activities and electrochemotherapy efficiency. J Mol Struct 2021; 1226: 129402. [http://dx.doi.org/10.1016/j.molstruc.2020.129402]

[108] Ruiz MC, Perelmulter K, LevA-n P, *et al.* Antiproliferative activity of two copper (II) complexes on colorectal cancer cell models: Impact on ROS production, apoptosis induction and NF-κB inhibition. Eur J Pharm Sci 2022; 169: 106092. [http://dx.doi.org/10.1016/j.ejps.2021.106092] [PMID: 34879254]

[109] López M, Tlahuext H, Pérez-Salgado M, *et al.* Synthesis, crystal structure, antibacterial, antiproliferative and QSAR studies of new bismuth(III) complexes of pyrrolidineditiocarbamate of dithia-bismolane and bismane, oxodithia- and trithia-bismocane. J Mol Struct 2020; 1217: 128456. [http://dx.doi.org/10.1016/j.molstruc.2020.128456]

[110] Sağlam EG, AkkoA S, Zeyrek CT, Dal H, Tutsak A-. New heterobimetallic nickel(II) ferrocenyldithiophosphonato complexes: Syntheses, characterization, antiproliferative activity and X-ray, DFT, molecular docking studies on trans-bis-[O-3-methyl-1-butyl-(ferrocenyl)dithiophosphonato]nickel(II). Inorg Chim Acta 2021; 514: 119991. [http://dx.doi.org/10.1016/j.ica.2020.119991]

[111] Carcelli M, Montalbano S, Rogolino D, *et al.* Antiproliferative activity of nickel(II), palladium(II) and zinc(II) thiosemicarbazone complexes. Inorg Chim Acta 2022; 533: 120779. [http://dx.doi.org/10.1016/j.ica.2021.120779]

[112] Amiri Rudbari H, Saadati A, Aryaeifar M, *et al.* Platinum(II) and Copper(II) complexes of asymmetric halogen-substituted [NNE1O] ligands: Synthesis, characterization, structural investigations and antiproliferative activity. Bioorg Chem 2022; 119: 105556. [http://dx.doi.org/10.1016/j.bioorg.2021.105556] [PMID: 34959175]

CHAPTER 5

Advances in the Synthesis of Heterocyclic Schiff Base Ligands with 3d-Transition Metals as Antibacterial Agents

Manisha[1] and **Pallavi Jain**[1,*]

[1] *Department of Chemistry, SRM Institute of Science and Technology, Delhi-NCR Campus, Modinagar, India*

Abstract: Heterocyclic Schiff Base Ligands (SBLs) and their 3d transition metal complexes have emerged as highly promising candidates in the field of antibacterial drug design. These compounds are synthesized through a condensation reaction between primary amines and aldehydes or ketones, resulting in an imine (-C=N-) functional group that serves as a versatile pharmacophore. The incorporation of heterocyclic moieties into Schiff bases further enhances their biological activity, while coordination with 3d transition metals such as Mn(II), Co(II), Cu(II), Zn(II), and Ni(II) forms metal complexes with superior stability and enhanced antibacterial activity. This chapter highlights recent advancements in the synthesis of heterocyclic SBLs with 3d transition metal complexes and examines their modes of interaction with bacterial strains. Through an analysis of both experimental and theoretical studies, we discuss the factors influencing the antibacterial activities of these metal complexes, including structural modifications, metal ion coordination, and ligand framework. The findings underscore the potential of heterocyclic Schiff base metal complexes as valuable agents in the development of novel antibacterial therapeutics.

Keywords: Antibacterial activity, Coordination chemistry, Geometry optimization, Schiff's base ligand, 3d-transition metal complex.

INTRODUCTION

Heterocyclic Schiff base ligands have become a focal point in the field of medicinal chemistry due to their versatile structural framework, ease of synthesis, and significant biological activities [1 - 3]. Schiff base ligands are typically synthesized by the condensation reaction of primary amines, aldehyde, and carbonyl compounds, resulting in compounds with an azomethine (-C=N-) group [4 - 6]. This functional group plays a crucial role in enhancing the reactivity and

* **Corresponding author Pallavi Jain:** Department of Chemistry, SRM Institute of Science and Technology, Delhi-NCR Campus, Modinagar, India; E-mail: palli24@gmail.com

Abhay Nanda Srivastva (Ed.)
All rights reserved-© 2026 Bentham Science Publishers

biological activity of Schiff bases, especially when coordinated with metal ions [7]. The incorporation of heterocyclic moieties into Schiff base ligands not only stabilizes these compounds but also amplifies their electronic and steric properties, making them particularly useful for antibacterial applications [8 - 10]. 3d transition metal complexes, particularly those with metals such as Mn(II), Co(II), Cu(II), Zn(II), and Ni(II), have been widely studied for their ability to enhance the pharmacological properties of Schiff bases [11 - 13] The metal ions interact with the azomethine group and other donor atoms in the Schiff base, forming stable complexes that often show superior antibacterial activity compared to the ligand alone [14 - 16]. The coordination with 3d transition metals allows these complexes to interact effectively with bacterial cell components, resulting in potential therapeutic applications against pathogenic bacteria, including both Gram-positive and Gram-negative strains [17, 18]. Advances in synthetic strategies have made it possible to develop heterocyclic Schiff base ligands with optimized structures and properties, enhancing their potential as antibacterial agents [19, 20]. Recent synthetic approaches focus on the careful selection of metal salts, ligands, and reaction conditions to achieve higher yields, better-defined structures, and antibacterial applications. The development of these Schiff base complexes offers promising avenues for new antibacterial agents that can combat infections more effectively than conventional treatments. This chapter explores recent progress in the synthesis of heterocyclic Schiff base ligands and their complexes with 3d transition metals, emphasizing the structural features that enhance their antibacterial properties. It further discusses the mechanisms by which these complexes exert their antibacterial effects, offering insights into their potential applications as therapeutic agents in the fight against bacterial infections.

Schiff's Base Approach to Antibacterial Activity

Bacteria can exist as single-celled organisms or form complex multicellular structures. Microscopic bacteria can initiate harmful processes that adversely affect human health and may develop resistance to chemical agents intended to control them [21]. Consequently, there is a pressing need to develop new antibacterial drugs.

The bidentate SBL 2-((1H-Benzoimidazole-4-ylimino)methyl)phenol **3** was synthesized through the condensation reaction between 2-aminobenzimidazole **1** and 4-hydroxybenzaldehyde **2**. Metal complexes of SBL**3** with Cu(II) **4**, Co(II) **5,** and Zn(II) **6** were also prepared (Scheme **1)** [22]. Elemental analysis and spectroscopic techniques, including Ultraviolet-Visible Spectroscopy (UV-Visible), Fourier-Transform Infrared Spectroscopy (FT-IR), CarbonNuclear Magnetic Resonance Spectroscopy (^{13}C-NMR), and Proton Nuclear Magnetic Resonance Spectroscopy (^{1}H-NMR), were synthesized to characterize compounds

3-6. The octahedral geometry of complexes **4-6** was confirmed. Cyclic voltammetry (CV) was utilized to examine the electrochemical properties of these complexes. The Antibacterialactivity of SBL **3** and theircomplexes Cu(II) **4**, Co(II) **5,** and Zn(II) **6** were evaluated against *Pseudomonas aeruginosa*(*P. aeruginosa), Escherichia coli (E. coli),* and *Staphylococcus aureus*(*S. aureus)*using the agar well diffusion method. The Cu(II) complex **4** showed the highest activity compared to other complexes.

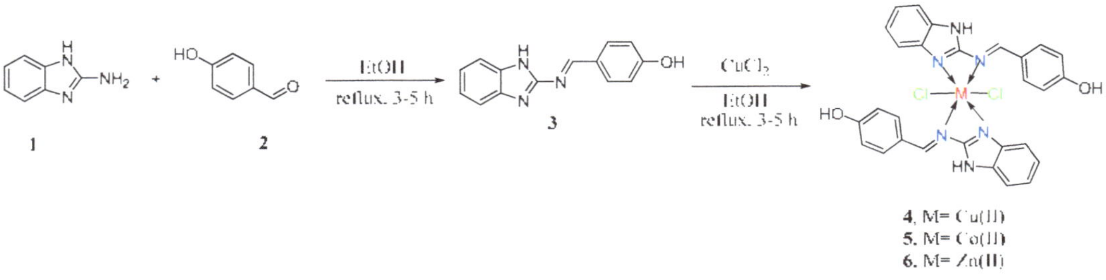

Scheme 1. The Synthesis of SBL **3** and its complexes **4-6**.

In 2020, S. Kumaresan *et al.* synthesized aSBL, 4-nitrobenzaldehydeglycylglycine **9**, by refluxing glycylglycine (GG) **8** with 4-nitrobenzaldehyde (4-NBA) **7**. Metal complexes of SBL**9** with Co(II) **10**, Cu(II) **11**, and Zn(II) **12** were also prepared (Scheme **2**) [23]. Characterization of compounds **9-12** was carried out using various spectroscopic methods (^{1}H-NMR, ^{13}C-NMR, FT-IR, UV-Visible, and XRD analysis. The tridentate nature of the SBL **9** was confirmed, showing coordination through azomethine nitrogen, carboxylate oxygen, and deprotonated peptide nitrogen atoms. All metal complexes were water-soluble. Spectroscopic analysis confirmed that the complexes exhibit octahedral geometry. The antibacterial activity of compounds **9-12** was assessed using the disc diffusion method against *E. coli, S. aureus*, *Bacillus subtilis (B. subtilis)*, and *P. aeruginosa*. The Cu(II) complex **11** showed the highest activity among the complexes.

In 2021, Hadi Kargar and his team synthesized Cu(II) complexes **19-22**fromfour SBL**15-18** (Scheme **3**) [24]. Compounds **15-22** were characterized using molar conductivity, elemental analysis, and spectroscopic techniques like FT-IR, UV-Visible, ^{1}H, and ^{13}C-NMR, and XRD analysis. All spectroscopic studies indicated a distorted tetrahedral geometry for complexes **19-22**. The antibacterial potential of **15-22** was evaluated against *S. aureus* and *E. coli*, bacterial strains. The **19-22** showed stronger antibacterial potential compared to **15-18**.

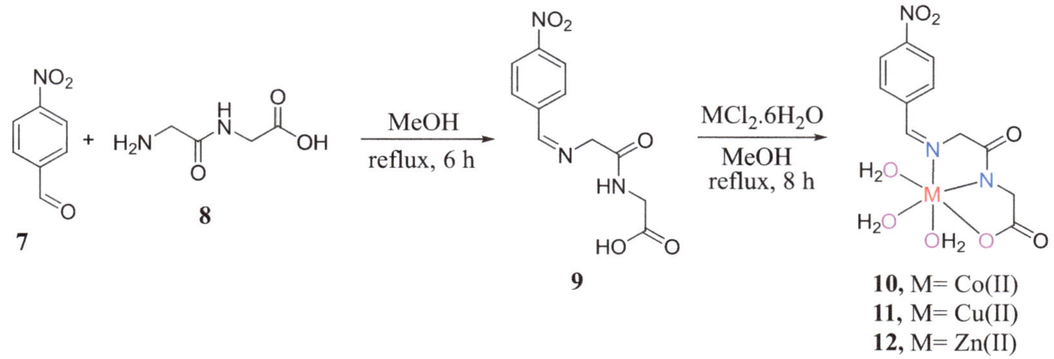

Scheme 2. Synthesis of SBL **9** and complexes**10-12**.

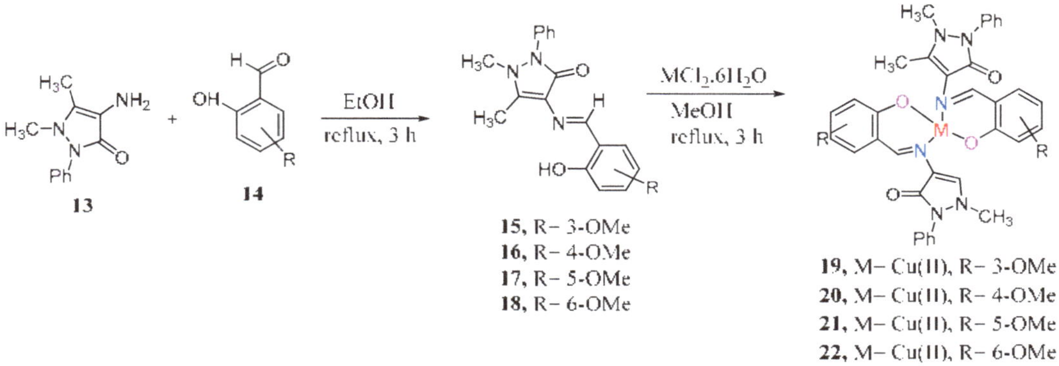

Scheme 3. The synthesis of SBL **15-18** and Cu(II) **19-22** complexes.

A pyrazole-based 'NO' bidentate SBL **25**, was synthesized from 3-amino-5-methylpyrazole **23** and 5-nitrosalicylaldehyde **24**. The Co(II) **28, 31**, Ni(II) **29, 32**, Cu(II) **26, 27**, and Zn(II) **30, 33** complexes of **25** were also prepared (Scheme 4) [25]. The SBL **25** and its metal complexes **26-33** were characterized using a spectroscopic method, such as FT-IR, ^{1}H NMR, mass, UV-Visible, TGA, cyclic voltammetry, magnetic susceptibility measurements, and elemental analysis. Spectral and analytical data confirmed a square planar geometry for complexes **26** and **27**, while complexes **28-33** were found to have octahedral geometry. The antibacterial potential of **25** and **26-33** was evaluated against food-borne pathogens, including *S. aureus, Clostridium perfringens (C. perfringens)*, *Listeria monocytogenes (L. monocytogenes)*, *E. coli* and *P. aeruginosa*. The results indicated that the Zn complexes **30, 33** exhibited the highest antibacterial activity compared to the other metal complexes.

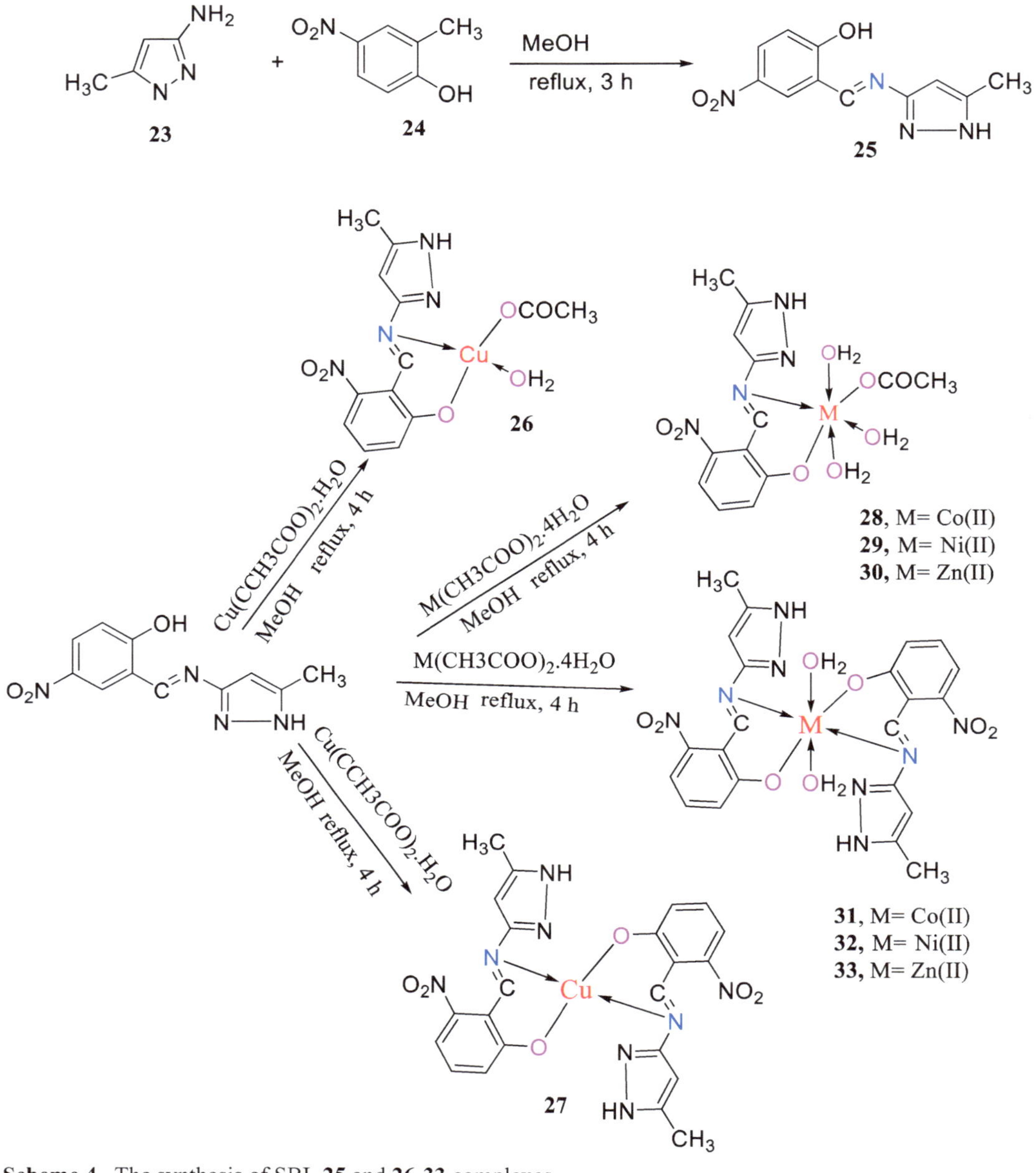

Scheme 4. The synthesis of SBL **25** and **26-33** complexes.

The synthesis, reactivity, structure, and antibacterial potential of (SBL) **36** were thoroughly examined through both experimental and theoretical methods, employing the B3LYP approach of Density Functional Theory (DFT). SBL **36** was synthesized *via* a reaction between salicylaldehyde **34** and amino alcohols **35**

(Scheme **5**) [26]. Physicochemical properties and various spectroscopic techniques were used to confirm the structures of the synthesized SBL **36**. Computational studies provided insights into molecular orbitals, chemical reactivity, stability, and molecular electrostatic potential. Thermodynamic parameters and optimized gaseous-state structures were also assessed using the B3LYP DFT method. The antibacterial activity of SBL **36** was tested against *P. aeruginosa* and *Bacillus cereus (B. cereus)* strains, demonstrating its potential as an antibacterial agent.

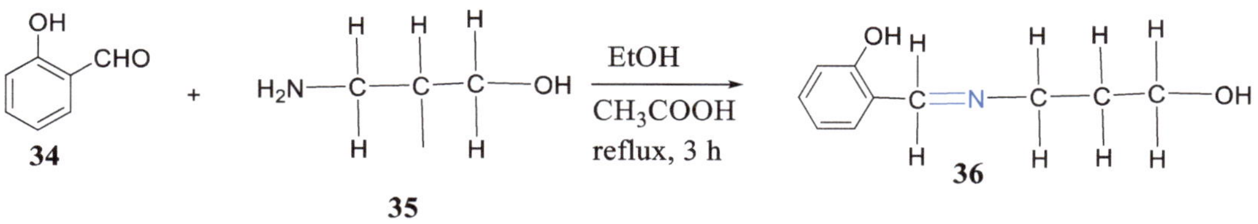

Scheme 5. The synthesis of SBL**36**.

The Mn(II) complex **40** was synthesized from the SBL **39** *via* condensation reaction of 2-(4-aminophenyl)ethan-1-ol **37** with ethyl alcohol **38** (Scheme **6)** [27]. The synthesized compounds were characterized using FT-IR, UV-Visible, mass spectrometry, ^{1}H, and ^{13}C-NMR, along with magnetic susceptibility and conductivity measurements. Spectroscopic and analytical data confirmed that SBL **39** acts as a monobasic bidentate ligand, forming an octahedral geometry forMn(II) complex **40**. The antibacterial activity of SBL **39** and the Mn(II) complex **40** was evaluated against*B. subtilis, S. aureus,* and *E. coli*, bacterial strains using the disc diffusion method. The complex 40 showed higher activity than SBL **39**.

A novel tetradentate SBL **43** was synthesized through a condensation reaction between 5-(2-fluorophenyl)furan-2-carbaldehyde **42** and benzene-1,2-diamine **41,** and its Zn(II) complex **44** (Scheme **7)** [28]. The SBL **43** and Zn(II) complex **44** was characterized using various spectroscopic methods, such as FT-IR, UV-Visible, XRD, TGA, and mass spectrometry. The *in vitro* antibacterial activity of the SBL**43** and its Zn(II) complex **44** was assessed against *S. pneumoniae, B. subtilis*, *P. aeruginosa, Salmonella typhi (S. typhi),* and *E. coli*. bacteria strain. The Zn(II) **44** demonstrated higher activity compared to the SBL **43**.

The SBL **47** was synthesized through a condensation reaction between 3,5-diiodosalicylaldehyde **46** and sulfamethazine **45**, and also Co(II) **48**, Cu(II) **49**, Ni(II) **50**, and Zn(II) **51**(Scheme **8)** [29]. The SBL **47** and its complexes **48-51** were characterized using spectroscopic techniques (FT-IR, UV-Visible, ^{1}H and ^{13}C-NMR), TGA, elemental analysis, magnetic susceptibility, and conductance measurements. Analytical and spectroscopic results confirmed that SBL **47** acts as a bidentate ligand, and octahedral geometries of all complexes **48-51**. The antibacterial potential of **47** and **48-51** was evaluated against *Enterococcus faecalis (E. faecalis), S. aureus, Proteus mirabilis (P. mirabilis)*, and *Klebsiella pneumoniae (K. pneumoniae)*. The Zn(II) **51** complex exhibited highest antibacterial activity compared to the other complexes.

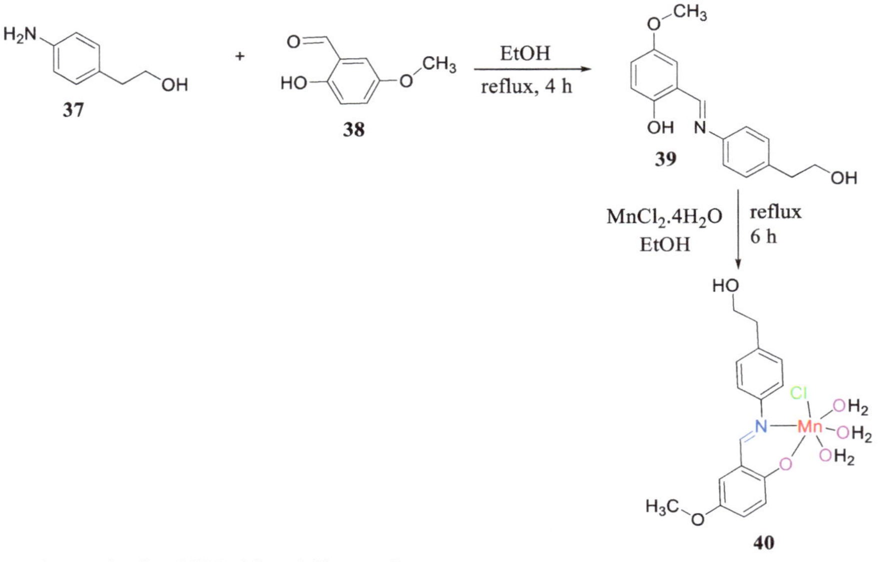

Scheme 6. The synthesis of SBL **39** and **40** complexes.

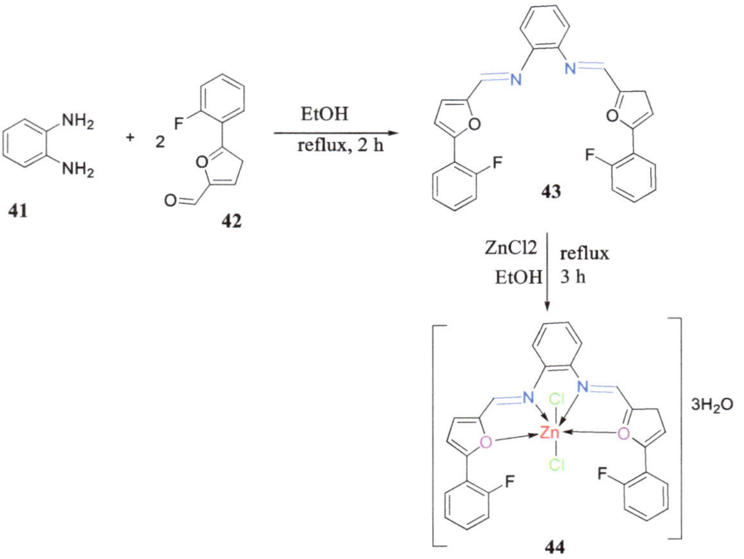

Scheme 7. The synthesis of SBL **43** and **44** complexes.

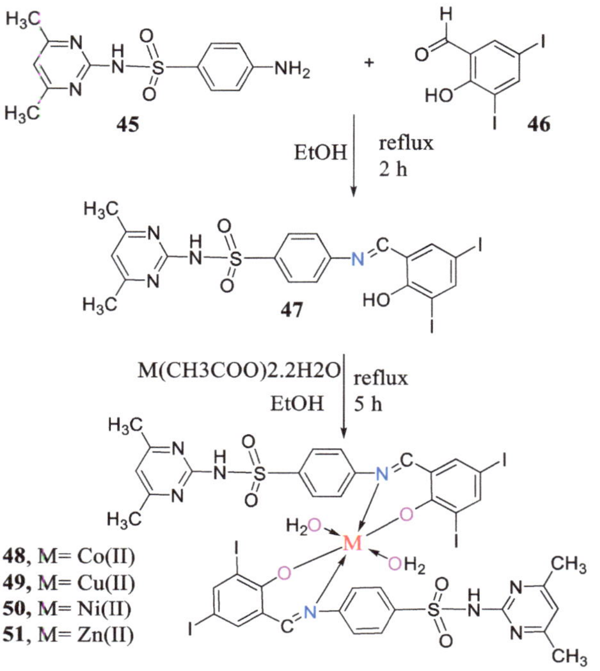

48, M= Co(II)
49, M= Cu(II)
50, M= Ni(II)
51, M= Zn(II)

Scheme 8. The synthesis of SBL **47** and **48-51** complexes.

A. A. Mohamed *et al.* synthesized SBL **54** through a condensation reaction between aniline **53** and salicylaldehyde **52**, along with its Co(II) **55** and Cu(II) **56** complexes, (Scheme **9)** [30]. The synthesized SBL **54** and its complexes were characterized using various FT-IR, ^{1}H NMR, TGA, and UV-Visible spectroscopy. All studies confirmed that SBL **54** acts as a bidentate ligand, while the Co(II) and Cu(II) complexes **55** and **56** exhibited an octahedral geometry. Antibacterial activity of SBL **54** and Co(II) **55** and Cu(II) **56** complexes was assessed against *S. aureus* and *Micrococcus luteus (M. luteus)*using the disc diffusion method. The complexes **56** exhibited enhanced activity compared to **55** and **54**.

A new mixed-ligand Ni(II) **62** complex [Ni: L1: L2] was synthesized from SBL **60** and 2-amino-4-methyl pyrimidine **61 (**Scheme **10)** [31]. The synthesized SBL **60** and Ni(II) **62** were characterized through the physicochemical techniques, including elemental analysis (CHN), conductometry, magnetic susceptibility measurements, FT-IR, ^{1}H NMR, TGA, and UV-Visible spectroscopy. The analysis confirmed the bidentate nature of **60**, while the spectroscopic data indicated an octahedral geometry for complex **62**. Both compounds **60** and **62** were assessed for their antibacterial potential against *B. subtilis*, *E. coli,* and *P. Aeruginosa* bacterial strains using the disc diffusion method. The complex **62** exhibits stronger antibacterial effects than SBL **60**.

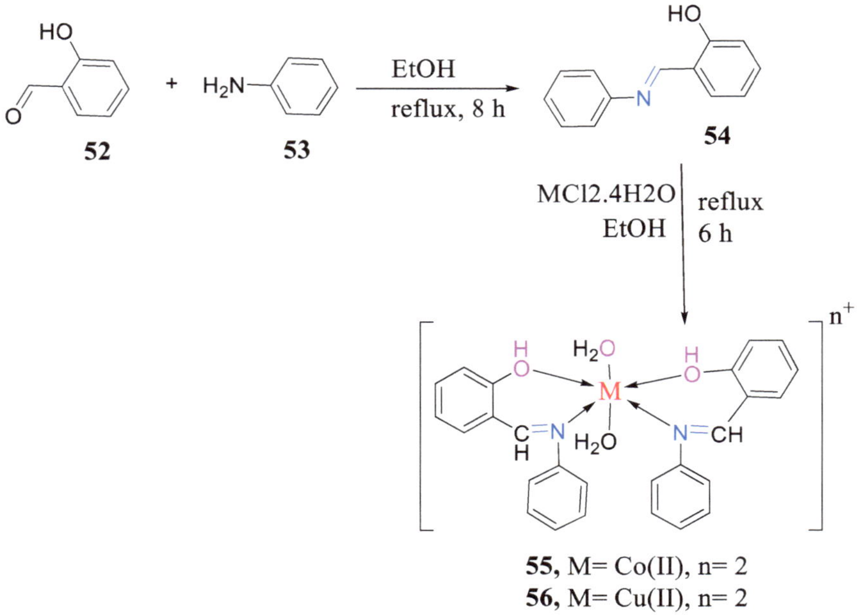

Scheme 9. The synthesis of SBL **54** and **55-56** complexes.

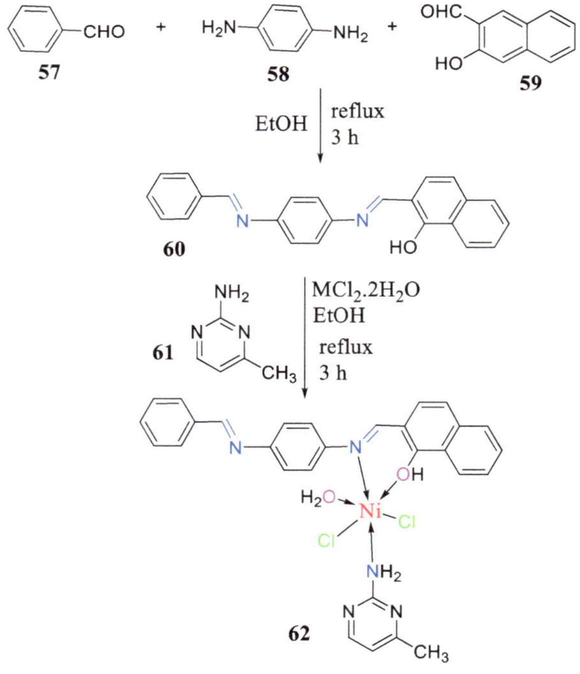

Scheme 10. The synthesis of SBL **60** and **62** complexes.

The tetradentate SBL **65** was synthesized *via* condensation of 2-hydroxy naphthaldehyde **63** with 2-amino benzhydrazide **64**, and Co(II) **66**, Cu(II) **67**, Ni(II) **68**, and Zn(II) **69** complexes (Scheme **11**) [32]. The **65** and **66-69** were characterized through different spectroscopic methods, including UV-Visible, FT-IR, ^{1}H-NMR, and mass spectrometry, as well as TGA, conductivity, and magnetic moment measurements. Analyses confirmed that SBL **65** acts as a tetradentate ligand, coordinating to metal ions through oxygen and nitrogen atoms, leading to octahedral geometries in complexes **66-69**. The antibacterial activity of **65** and **66-69** was evaluated against *K. pneumoniae* and *S. aureus* using the agar well diffusion method. The findings reveal that the Cu(II) complex **67** demonstrated the strongest activity against *K. pneumoniae*, whereas the Zn(II) complex **69** exhibited the highest activity against *S. aureus*.

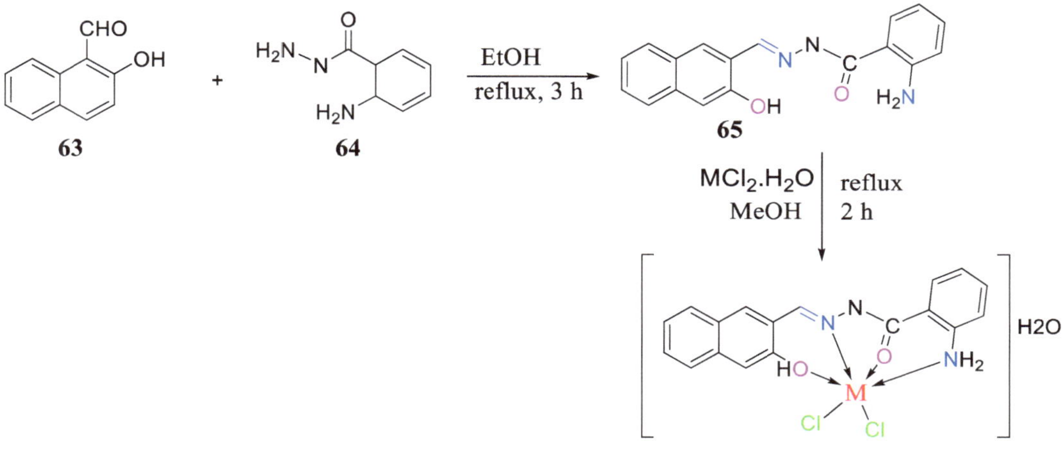

66, M= Co(II)
67, M= Cu(II)
68, M= Ni(II)
69, M= Zn(II)

Scheme 11. The synthesis of SBL **65** and **66-69** complexes.

A novel SBL **72** was synthesized through the condensation reaction of 3-amino-5-methylpyrazole **70** with 5-bromosalicylaldehyde **71** in a 1:1 ratio, along with its Co(II) **75, 78**, Cu(II) **73, 74**, Ni(II) **76, 79**, and Zn(II) **77, 80** complexes (Scheme **12**) [33]. The SBL **72** and complexes **73-80** were characterized by elemental analysis and different spectroscopic methods, including ^{1}H-NMR, FT-IR, UV-Visible, and mass spectrometry. Complexes **75-80** were found to have an octahedral geometry, while complexes **73** and **74** displayed a square planar structure. The antibacterial activity of SBL **72** and complexes **73-80** was evaluated against *S. aureus, C. perfringens, L. monocytogenes, E. coli,* and *P.*

aeruginosa. All complexes exhibited antibacterial activity, whereas Zn(II) **77** and 80 complexes showed the highest activity (MIC of 6.25 µg/mL) across all tested bacterial strains, comparable to the standard drug.

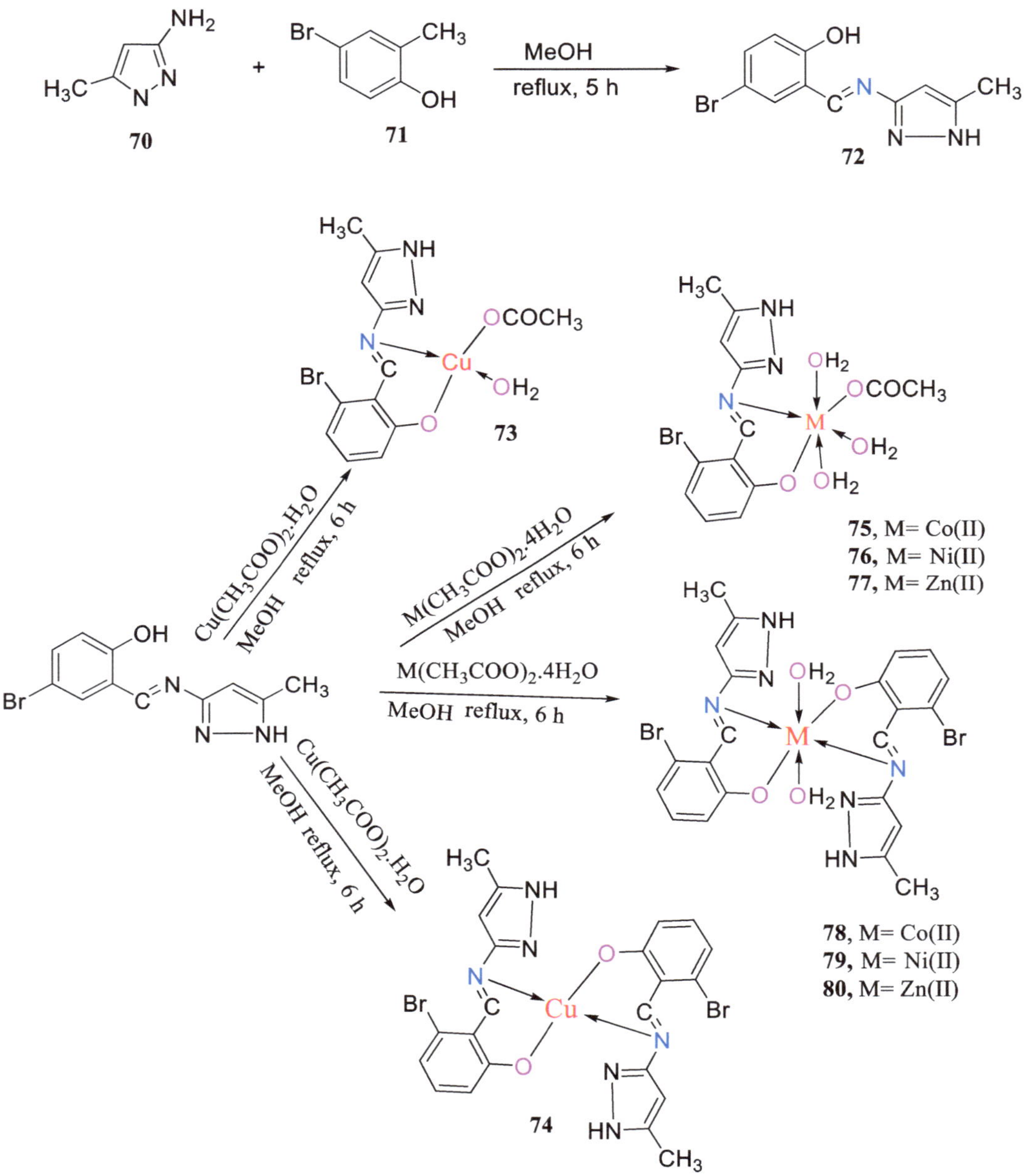

Scheme 12. The synthesis of SBL **72** and **73-80** complexes.

S. Saleem *et al.* investigated the synthesis of a novel SBL, ((o-tolylimino)methyl)phenol **83**, by the condensation reaction of 2-hydroxybenzaldehyde **81**with o-toluidine**82** along with its Mn(II) **84**, Co(II) **85**, Ni(II) **86**, and Cu(II) **87** complexes (Scheme **13**) [34]. The characterization of these compounds was carried out using different spectroscopic methods, including FT-IR, ^{1}H NMR, UV-Visible, TGA, XRD, mass spectrometry, and magnetic measurements. The spectroscopic data revealed that the Mn(II) **84** and Co(II) **85** complexes have an octahedral geometry, whereas the Cu(II) 87 complex has a square-planar structure. The antibacterial potential of the SBL **83** and its Mn(II) **84** and Co(II) **85** complexes was evaluated against bacterial strains *E. coli, K. pneumoniae, S. aureus, and B. subtilis* using the disc diffusion method. The Ni(II) **86** complex showed the highest antibacterial activity against *E. coli and K. pneumoniae*, compared to the other complexes.

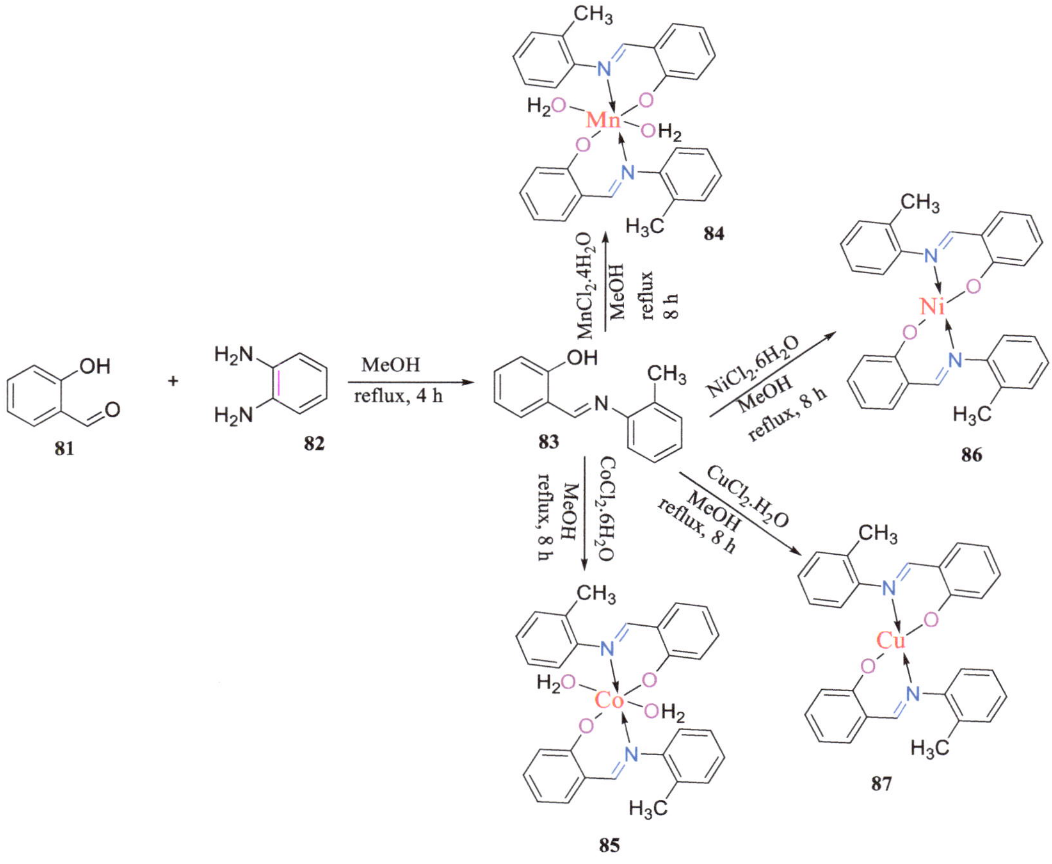

Scheme 13. The synthesis of SBL **83** and **84-87** complexes.

Novel Cu(II) **91** and Zn(II) **92** complexes were synthesized from the SBL **90** (Scheme **14)** [35]. The structures of SBL **90** and its metal complexes were confirmed through elemental analysis, UV-Visible spectroscopy, FT-IR, ^{1}H-NMR, ESR, and TGA, supporting the proposed geometries of the complexes. ESR and UV-Visible studies revealed that both Cu(II) **91** and Zn(II) **92** complexes exhibited a square planar geometry. The *in vitro* antibacterial activity of the synthesized compounds **90-92** was tested against five bacterial strains: *Bacillus amyloliquefaciens (B. amyloliquefaciens), Staphylococcus epidermidis (S. epidermidis), S. aureus, B. cereus*, and *B. subtilis* using the disc diffusion method in agar media. Among the tested compounds, the Cu(II) complex **91** showed the highest antibacterial potential against *S. aureus* compared to the Zn(II) complex **92**.

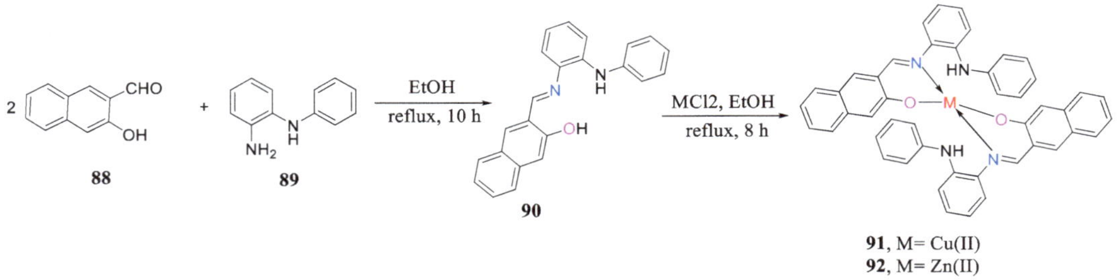

Scheme 14. The synthesis of SBL **90** and **91-92** complexes.

A new series of Co(II) **97** and Cu(II) **98** complexes was synthesized using an imine SBL **95**, derived from 4-methyl-o-phenylenediamine **93**, furfural **94**, and 2,2'-bipyridine **96** (Scheme **15**) [36]. The structures of these complexes were analyzed using FT-IR, ^{1}H-NMR, UV-Visible, mass spectrometry, elemental analysis, TGA, conductivity, and magnetic susceptibility measurements. All spectroscopic studies confirmed octahedral geometries for the Co(II) 97 and Cu(II) **98** complexes. Conductivity measurements indicated that all complexes display electrolytic behaviour. The antibacterial activity of SBL **95** and its Co(II) **97** and Cu(II) **98** complexes was confirmed against *S. aureus, B. subtilis, P. vulgaris*, and *E. coli* using the paper disk diffusion method. The complexes 97 and 98 exhibited stronger antibacterial activity than **95**.

Mn(II) **102** and Co(II) **103** complexes were synthesized from the SBL **101**, using a 2:1 ligand-to-metal salt ratio (Scheme **16**) [37]. The **101, 102,** and **103** were analyzed using various spectroscopic methods such as FT-IR, UV-Visible, and EPR. Elemental and spectroscopic studies confirmed the ligand's bidentate donor nature and an octahedral geometry for both complexes. Conductivity data indicated that both **102** and **103** complexes are non-electrolytic. *In vitro* antibacterial screening was conducted using the agar well diffusion method against *E. coli* and *S. aureus*. The Co(II) **103** complex exhibited stronger antibacterial activity compared to the Mn(II) **102** complex.

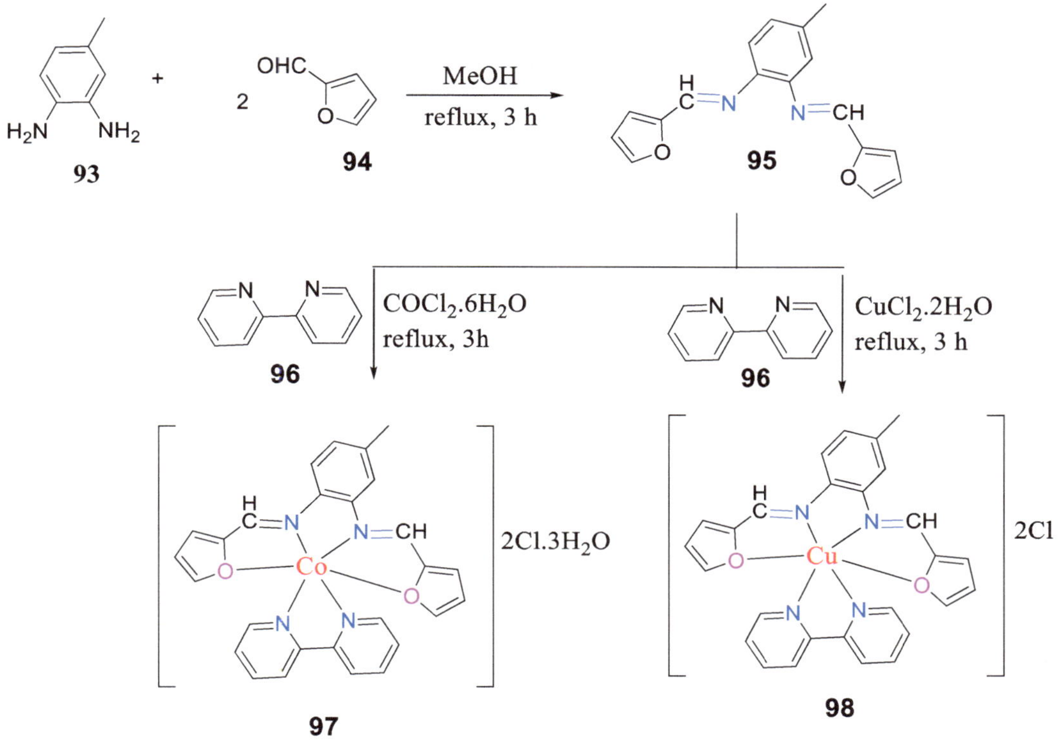

Scheme 15. The synthesis of SBL **95** and **97-98** complexes.

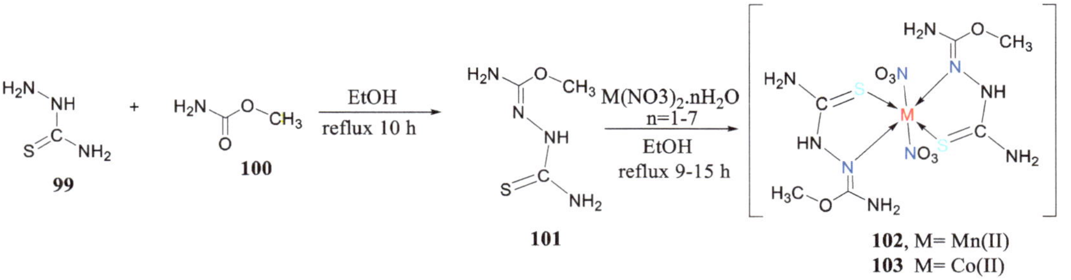

Scheme 16. The synthesis of SBL **101** and **102-103** complexes.

The SBL **106** was synthesized by refluxing 2-aminopyridine-3-thiol **104** with benzil **105** in a 2:1 molar ratio. Using a similar method, Mn(II)**107**, Co(II)**108**, Ni(II)**109**, and Cu(II) **110** complexes were also prepared (Scheme **17**) [38]. Characterization of all compounds was conducted through various analytical and spectroscopic techniques, including melting point, molar conductance, magnetic moment, ^{1}H-NMR, FT-IR, mass, EPR, and UV-Visible spectroscopy. The antibacterial potential of **106** and **107-110** was evaluated against *B. subtilis* and *S. typhi* using the well-diffusion method, with Gentamycin and Amphotericin B as standard drugs. The complex Ni(II)**109** exhibited stronger antibacterial activity than other complexes.

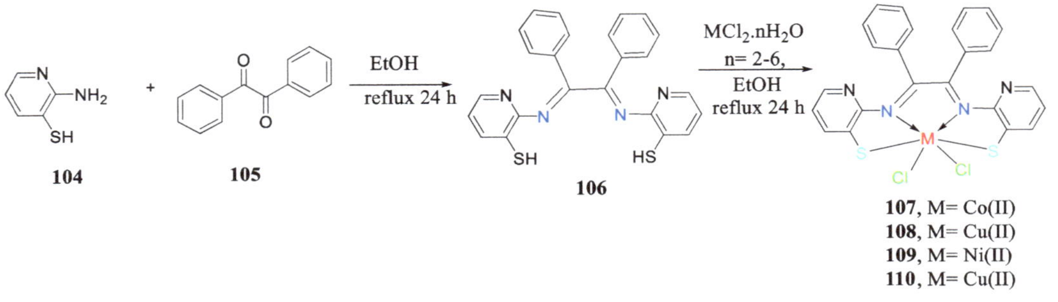

Scheme 17. The synthesis of SBL **106** and **107-110** complexes.

Mechanistic Insights into the Superior Activity of Cu(II) and Zn(II) Complexes

Among the synthesized complexes, Cu(II) and Zn(II) derivatives consistently demonstrated superior antimicrobial and antifungal activity compared to other metal ions. This enhanced performance can be rationalized by considering several physicochemical and biological factors:

Coordination Geometry and Flexibility

Cu(II) complexes exhibit variable coordination numbers and geometries (square planar, square pyramidal, or octahedral), which allow them to adapt effectively to biological targets. Such structural flexibility facilitates stronger interactions with microbial enzymes, proteins, and nucleic acids. Zn(II) complexes, on the other hand, often stabilize in tetrahedral or octahedral geometries that provide favorable binding orientations with biomolecules.

Redox Properties and ROS Generation

Cu(II) is a redox-active metal capable of cycling between Cu(II)/Cu(I) states. This redox cycling can lead to the *in-situ* generation of Reactive Oxygen Species (ROS), such as hydroxyl and superoxide radicals, which induce oxidative stress and damage microbial membranes, proteins, and DNA. Zn(II), though redox-inactive, enhances the stability of the ligand framework and can indirectly support ROS-mediated toxicity in the biological environment.

Lipophilicity and Membrane Permeation

Chelation with Schiff base ligands reduces the polarity of the metal ion through partial sharing of the ligand donor atoms. This increases the overall lipophilicity of the complex, improving its ability to permeate microbial lipid membranes. Cu(II) and Zn(II) complexes generally display higher partitioning into lipid bilayers, which enhances their uptake and activity inside microbial cells.

Possible Modes of Action

DNA/BSA Interaction: Cu(II) complexes show strong binding affinity toward DNA bases, potentially intercalating or cleaving DNA strands through ROS mediation. Zn(II) complexes stabilize interactions with nucleic acids and proteins *via* Lewis acid–base coordination.

Enzyme Inhibition: Both Cu(II) and Zn(II) ions are known to inhibit metalloenzymes essential for microbial survival.

Cell Wall/Membrane Disruption: Increased lipophilicity enables penetration through cell walls, leading to leakage of intracellular contents and microbial cell death.

Taken together, these features explain why Cu(II) and Zn(II) complexes consistently surpass other metal derivatives in biological activity. The synergy of favorable geometry, redox behavior (for Cu), Lewis acidity (for Zn), and enhanced lipophilicity accounts for their potent antimicrobial and antifungal effects.

Limitations and Future Perspectives in Drug Development

Although Schiff base metal complexes have demonstrated promising antimicrobial and antifungal activities, translating these compounds into clinically useful drugs is not without challenges. One of the foremost issues is **solubility**, as many of these complexes are poorly soluble in aqueous media, which restricts their absorption and bioavailability in biological systems. Limited solubility often reduces therapeutic effectiveness and hinders their potential as systemic drugs.

Another major concern is **bioavailability**. Even when soluble, the complexes may undergo rapid metabolism or clearance from the body, limiting the concentration that reaches the target site. Moreover, the presence of the metal center raises possible **toxicity concerns**, as high doses or long-term administration could lead to undesirable side effects or accumulation in tissues. These limitations highlight the gap between laboratory efficacy and clinical applicability.

To overcome these challenges, several **drug delivery strategies** are being explored. **Nanoformulations**, such as nanoparticles, liposomes, or solid lipid carriers, can improve solubility, enhance membrane permeability, and allow controlled drug release. Similarly, **polymer conjugates** can stabilize the complexes, reduce toxicity, and provide sustained delivery. In addition, **targeted delivery systems**, including ligand-functionalized nanoparticles, can direct the complexes specifically to infected or cancerous tissues, thereby maximizing therapeutic efficacy while minimizing side effects.

Overall, while Schiff base metal complexes hold significant promise as next-generation antimicrobial agents, their successful clinical translation will depend on addressing solubility, bioavailability, and toxicity issues through advanced formulation and delivery strategies.

Table 1. Comparative summary.

Ligand (SBL)	Metal Complexes	Geometry	Key Pathogens Tested	Most Active Complex
SBL 3 (Bidentate benzimidazole-based)	Cu(II) 4, Co(II) 5, Zn(II) 6	Octahedral	*P. aeruginosa, E. coli, S. aureus*	Cu(II) 4
SBL 9 (Tridentate peptide-based)	Co(II) 10, Cu(II) 11, Zn(II) 12	Octahedral	*E. coli, S. aureus, B. subtilis, P. aeruginosa*	Cu(II) 11
SBL 15–18	Cu(II) 19–22	Distorted tetrahedral	*S. aureus, E. coli*	Cu(II) complexes (19–22)
Pyrazole SBL 25	Co(II) 28, 31; Ni(II) 29, 32; Cu(II) 26, 27; Zn(II) 30, 33	Square planar / Octahedral	*S. aureus, C. perfringens, L. monocytogenes, E. coli, P. aeruginosa*	Zn(II) 30, 33
SBL 36 (Salicylaldehyde–amino alcohol)	–	–	*P. aeruginosa, B. cereus*	SBL 36 itself
SBL 39 (Monobasic bidentate)	Mn(II) 40	Octahedral	*B. subtilis, S. aureus, E. coli*	Mn(II) 40
SBL 43 (Tetradentate)	Zn(II) 44	Octahedral	*S. pneumoniae, B. subtilis, P. aeruginosa, S. typhi, E. coli*	Zn(II) 44
SBL 47 (Bidentate)	Co(II) 48, Cu(II) 49, Ni(II) 50, Zn(II) 51	Octahedral	*E. faecalis, S. aureus, P. mirabilis, K. pneumoniae*	Zn(II) 51
SBL 54 (Bidentate)	Co(II) 55, Cu(II) 56	Octahedral	*S. aureus, M. luteus*	Cu(II) 56

(Table 1) cont.....

Ligand (SBL)	Metal Complexes	Geometry	Key Pathogens Tested	Most Active Complex
SBL 60 (Bidentate)	Ni(II) 62 (mixed ligand)	Octahedral	*B. subtilis, E. coli, P. aeruginosa*	Ni(II) 62
SBL 65 (Tetradentate)	Co(II) 66, Cu(II) 67, Ni(II) 68, Zn(II) 69	Octahedral	*K. pneumoniae, S. aureus*	Cu(II) 67 (K. pneumoniae), Zn(II) 69 (*S. aureus*)
SBL 72 (Pyrazole-based)	Co(II) 75, 78; Cu(II) 73, 74; Ni(II) 76, 79; Zn(II) 77, 80	Square planar / Octahedral	*S. aureus, C. perfringens, L. monocytogenes, E. coli, P. aeruginosa*	Zn(II) 77, 80 (MIC 6.25 µg/mL)
SBL 83 (Tolyl-based)	Mn(II) 84, Co(II) 85, Ni(II) 86, Cu(II) 87	Octahedral / Square planar	*E. coli, K. pneumoniae, S. aureus, B. subtilis*	Ni(II) 86
SBL 90 (Phenolic-imine)	Cu(II) 91, Zn(II) 92	Square planar	*B. amyloliquefaciens, S. epidermidis, S. aureus, B. cereus, B. subtilis*	Cu(II) 91
SBL 95 (Imine-bipyridine)	Co(II) 97, Cu(II) 98	Octahedral	*S. aureus, B. subtilis, P. vulgaris, E. coli*	Both Co(II) 97 & Cu(II) 98
SBL 101 (Bidentate)	Mn(II) 102, Co(II) 103	Octahedral	*E. coli, S. aureus*	Co(II) 103
SBL 106 (Thiol-pyridine)	Mn(II) 107, Co(II) 108, Ni(II) 109, Cu(II) 110	Octahedral	*B. subtilis, S. typhi*	Ni(II) 109

CONCLUSION

Transition metal complexes derived from SBL have garnered significant attention for their antibacterial properties. This study concludes that these metal complexes exhibit superior antibacterial activity compared to SBL alone. The enhanced efficacy is primarily attributed to the delocalization of π-electrons across the aromatic rings during chelation, which increases the lipophilicity of the metal complexes. Additionally, various factors contribute to their improved biological activity, including the type of donor atoms or metal ions, complex geometry, overall charge, solubility, conductivity, and the metal-ligand bond length. Complexes containing heterocyclic moieties and hydrophobic groups tend to exhibit stronger bioactivity, while the presence of electron-withdrawing substituents further amplifies their biological performance. This chapter underscores the potential of SBL and its metal complexes as biologically significant precursors in advanced research. Future studies should aim to develop new Schiff base metal complexes with specific structural features and to better understand how they work biologically.

REFERENCES

[1] Thakur S, Bhalla A. Sustainable synthetic endeavors of pharmaceutically active Schiff bases and their metal complexes: A review on recent reports. Tetrahedron 2024; 153: 133836.
[http://dx.doi.org/10.1016/j.tet.2024.133836]

[2] Jain P, Vishvakarma VK, Singh P, *et al.* Bioactive Thiosemicarbazone Coordination Metal Complexes: Synthesis, Characterization, Theoretical analysis, Biological Activity, Molecular Docking and ADME analysis. Chem Biodivers 2023; 20(8): e202300760.
[http://dx.doi.org/10.1002/cbdv.202300760] [PMID: 37427893]

[3] Younus HA, Saleem F, Hameed A, *et al.* Part-II: an update of Schiff bases synthesis and applications in medicinal chemistry-a patent review (2016-2023). Expert Opin Ther Pat 2023; 33(12): 841-64.
[http://dx.doi.org/10.1080/13543776.2023.2297729] [PMID: 38115554]

[4] Al-Mosawy M. Review of the biological effects of Schiff bases and their derivatives, including their synthesis. Medical Science Journal for Advance Research 2023; 4(2): 67-85.
[http://dx.doi.org/10.46966/msjar.v4i2.117]

[5] Jain P, Sharma S, Kumar N, Misra N. Ni(II) and Cu(II) complexes of bidentate thiosemicarbazone ligand: Synthesis, structural, theoretical, biological studies and molecular modeling. Appl Organomet Chem 2020; 34(9): e5736.
[http://dx.doi.org/10.1002/aoc.5736]

[6] Boulechfar C, Ferkous H, Delimi A, *et al.* Schiff bases and their metal Complexes: A review on the history, synthesis, and applications. Inorg Chem Commun 2023; 150: 110451.
[http://dx.doi.org/10.1016/j.inoche.2023.110451]

[7] Kumar M, Singh AK, Singh VK, Yadav RK, Singh AP, Singh S. Recent developments in the biological activities of 3d-metal complexes with salicylaldehyde-based N, O-donor Schiff base ligands. Coord Chem Rev 2024; 505: 215663.
[http://dx.doi.org/10.1016/j.ccr.2024.215663]

[8] Juyal VK, Pathak A, Panwar M, *et al.* Schiff base metal complexes as a versatile catalyst: A review. J Organomet Chem 2023; 999: 122825.
[http://dx.doi.org/10.1016/j.jorganchem.2023.122825]

[9] Alfonso-Herrera LA, Hernández-Romero D, Cruz-Navarro JA, *et al.* Transition metal complexes with tetradentate Schiff bases (N_2O_2) obtained from salicylaldehyde: A review of their possible anticancer properties. Coord Chem Rev 2024; 505: 215698.
[http://dx.doi.org/10.1016/j.ccr.2024.215698]

[10] Sanjurani T, Barman P. The versatile nature of indole containing Schiff bases: An Overview. Polyhedron 2024; 249: 116779.
[http://dx.doi.org/10.1016/j.poly.2023.116779]

[11] Kumari P, Choudhary M, Kumar A, *et al.* Copper(II) Schiff base complexes: Synthetic and medicinal perspective. Inorg Chem Commun 2023; 158: 111409.
[http://dx.doi.org/10.1016/j.inoche.2023.111409]

[12] Dasmahapatra U, Maiti B, Alam MM, Chanda K. Anti-cancer property and DNA binding interaction of first row transition metal complexes: A decade update. Eur J Med Chem 2024; 275: 116603.
[http://dx.doi.org/10.1016/j.ejmech.2024.116603] [PMID: 38936150]

[13] Manisha , Jain P, Yadav SK. Anticancer and antimicrobial activities of schiff base ligand and their Co(II), Cu(II), and Zn(II) transition metal complexes: An overview. Springer Proceedings in Materials 2024; 51: 257-98.
[http://dx.doi.org/10.1007/978-981-97-5169-3_19]

[14] Zabiszak M, Frymark J, Ogawa K, *et al.* Complexes of β-lactam antibiotics and their Schiff-base derivatives as a weapon in the fight against bacterial resistance. Coord Chem Rev 2023; 493: 215326. [http://dx.doi.org/10.1016/j.ccr.2023.215326]

[15] Gosu NR, Hazarathaiah Yadav C, Ramakrishna Reddy K, Losetty V, Kothinti RR, Sampath S. Investigating spectroscopic properties, biological effects, and molecular docking studies of synthesized Schiff base metal complexes. Polyhedron 2024; 255: 116954. [http://dx.doi.org/10.1016/j.poly.2024.116954]

[16] Jain P, Kumar D, Chandra S, Misra N. Experimental and theoretical studies of Mn(II) and Co(II) metal complexes of a tridentate Schiff's base ligand and their biological activities. Appl Organomet Chem 2020; 34(2): e5371. [http://dx.doi.org/10.1002/aoc.5371]

[17] Das B, Gupta P. Multimetallic transition metal complexes:Luminescent probes for biomolecule sensing, ion detection, imaging and therapeutic application. Coord Chem Rev 2024; 504: 215656. [http://dx.doi.org/10.1016/j.ccr.2024.215656]

[18] Weng C, Tan YLK, Koh WG, Ang WH. Harnessing transition metal scaffolds for targeted antibacterial therapy. Angew Chem Int Ed 2023; 62(50): e202310040. [http://dx.doi.org/10.1002/anie.202310040] [PMID: 37621226]

[19] Kumar M, Singh AK, Singh AK, *et al.* Recent advances in 3d-block metal complexes with bi, tri, and tetradentate Schiff base ligands derived from salicylaldehyde and its derivatives: Synthesis, characterization and applications. Coord Chem Rev 2023; 488: 215176. [http://dx.doi.org/10.1016/j.ccr.2023.215176]

[20] Kumar KBC, Yogesh GB, Basavaraju YB. Design, synthesis, and evaluation of triazoline schiff bases as potential antibacterial agents against methicillin-resistant *Staphylococcus aureus* (MRSA): A comprehensive computational and experimental study. J Mol Struct 2025; 1322: 140349. [http://dx.doi.org/10.1016/j.molstruc.2024.140349]

[21] Mukherjee S, Ghorai S. Fungal biology. Curr Dev Biotechnol Bioeng 2023; 0: 67-104. [http://dx.doi.org/10.1016/B978-0-323-91872-5.00017-X]

[22] Kalarani R, Sankarganesh M, Kumar GGV, Kalanithi M. Synthesis, spectral, DFT calculation, sensor, antimicrobial and DNA binding studies of Co(II), Cu(II) and Zn(II) metal complexes with 2-amino benzimidazole Schiff base. J Mol Struct 2020; 1206: 127725. [http://dx.doi.org/10.1016/j.molstruc.2020.127725]

[23] Shiju C, Arish D, Kumaresan S. Novel water soluble Schiff base metal complexes: Synthesis, characterization, antimicrobial-, DNA cleavage, and anticancer activity. J Mol Struct 2020; 1221: 128770. [http://dx.doi.org/10.1016/j.molstruc.2020.128770]

[24] Kargar H, Aghaei-Meybodi F, Behjatmanesh-Ardakani R, *et al.* Synthesis, crystal structure, theoretical calculation, spectroscopic and antibacterial activity studies of copper(II) complexes bearing bidentate schiff base ligands derived from 4-aminoantipyrine: Influence of substitutions on antibacterial activity. J Mol Struct 2021; 1230: 129908. [http://dx.doi.org/10.1016/j.molstruc.2021.129908]

[25] Devi P, Singh K, Kumar B, Singh JK. Synthesis, spectroscopic, antimicrobial and *in vitro* anticancer activity of Co^{+2}, Ni^{+2}, Cu^{+2} and Zn^{+2} metal complexes with novel Schiff base. Inorg Chem Commun 2023; 152: 110674. [http://dx.doi.org/10.1016/j.inoche.2023.110674]

[26] DeepikaVerma , Sharma S, Vashishtha M. Evaluation of optimized molecular structure-antimicrobial and antioxidant efficacy relationship of Schiff bases. Environ Sci Pollut Res Int 2022; 30(8): 20874-86. [http://dx.doi.org/10.1007/s11356-022-23633-1] [PMID: 36260232]

[27] Durairaj P, Maruthavanan T, Manjunathan S, Subashini S, Rokhum SL, Baskar G. Microwave assisted synthesis, characterization and bioactivity evaluation of a cobalt (II) complex with a novel Schiff base ligand derived from phenylacetyl urea and salicylaldehyde. J Mol Struct 2024; 1295: 136650. [http://dx.doi.org/10.1016/j.molstruc.2023.136650]

[28] Amri N, Alaghaz ANMA. Synthesis, structural investigations, *in vitro* cytotoxicity, apoptotic activity, cell cycle analysis, and molecular modeling studies of nano-sized Zn(II) Schiff base complex. Appl Organomet Chem 2024; 38(8): e7577. [http://dx.doi.org/10.1002/aoc.7577]

[29] Saeed M, Parveen B, Rasool N, Ghulam Ali K, Munawar KS. Synthesis, characterization, *in vitro* and *in silico* biological studies of 3d transition metal(II) chelates of a Schiff base derived from sulfamethazine and 3,5-diiodosalicylaldehyde. J Coord Chem 2024; 77(15-16): 1860-86. [http://dx.doi.org/10.1080/00958972.2024.2387760]

[30] Mohamed AA, Nassr AA, Sadeek SA, Rashid NG, Abd El-Hamid SM. First report on several NO-donor sets and bidentate schiff base and its metal complexes: Characterization and antimicrobial investigation. Compounds 2023; 3(3): 376-89. [http://dx.doi.org/10.3390/compounds3030029]

[31] Al-Fakeh MS, Alsikhan MA, Alnawmasi JSH, Alluhayb AH, Al-Wahibi MS. New nanosized V(III), Fe(III), and Ni(II) complexes comprising schiff base and 2-amino-4-methyl pyrimidine: Synthesis, properties, and biological activity. Int J Biomater 2024; 2024: 1-13. [http://dx.doi.org/10.1155/2024/9198129] [PMID: 39262654]

[32] Mohan B, Shaalan N. Synthesis, spectroscopic, and biological activity study for new complexes of some metal ions with schiff bases derived from 2-hydroxy naphthaldehyde with 2-amine benzhydrazide. Ibn AL-Haitham Journal For Pure and Applied Sciences 2023; 36(1): 208-24. [http://dx.doi.org/10.30526/36.1.2978]

[33] Devi P, Singh K, Kubavat B. Synthesis, spectroscopic, quantum, thermal and kinetics, antibacterial and antifungal studies: Novel Schiff base 5-methyl-3-((5-bromosalicylidene) amino)- pyrazole and its transition metal complexes. Results Chem 2023; 5: 100813. [http://dx.doi.org/10.1016/j.rechem.2023.100813]

[34] Saleem S, Parveen B, Abbas K, Iqbal S, Altaf AA, Kausar S. Synthesis, structural elucidation, molecular modeling and antimicrobial studies of Mn(II), Co(II), Ni(II), and Cu(II) complexes containing NO donor bidentate Schiff base. Appl Organomet Chem 2023; 37(10): e7234. [http://dx.doi.org/10.1002/aoc.7234]

[35] Gurusamy S, Krishnaveni K, Sankarganesh M, Nandini Asha R, Mathavan A. Synthesis, characterization, DNA interaction, BSA/HSA binding activities of VO(IV), Cu(II) and Zn(II) Schiff base complexes and its molecular docking with biomolecules. J Mol Liq 2022; 345: 117045. [http://dx.doi.org/10.1016/j.molliq.2021.117045]

[36] Ismail BA, Abd El-Wahab ZH, Ali OAM, Nassar DA. Synthesis, structural characterization, and antimicrobial evaluation of new mononuclear mixed ligand complexes based on furfural-type imine ligand, and 2,2′-bipyridine. Sci Rep 2023; 13(1): 9196. [http://dx.doi.org/10.1038/s41598-023-36060-0] [PMID: 37280267]

[37] Alka , Vishvakarma VK, Yadav S, Singh P, Jain P. *In Vitro*, *In Silico*, ADME and Theoretical Analysis of Mn(II) and Co(II) Complexes Derived from Methyl-(*Z)-N* '-Carbamothioylcarbamohydrazonate Schiff Base Ligand. Chem Biodivers 2023; 20(6): e202300042. [http://dx.doi.org/10.1002/cbdv.202300042]

[38] Alka , Singh J, Kumari P, Jain P. Synthesis, characterization, biological, ADMET, and molecular docking studies of transition metal complexes of aminopyridine schiff base derivative. Chem Biodivers 2024; 21(10): e202401101. [http://dx.doi.org/10.1002/cbdv.202401101]

CHAPTER 6

Ferrocene-Based Hybrids as Potent Anticancer Agents: Recent Advances and Structure-Activity Insights

Anindra Sharma[1,*]

[1] *Department of Chemistry, Lalit Narayan Mithila University, Darbhanga, Bihar, India*

Abstract: Cancer remains a leading global health concern, second only to cardiovascular diseases. A significant challenge in managing and eliminating cancer is the continual emergence of drug-resistant strains and the limited selectivity of current anticancer therapies. This underscores the urgent need to develop novel and more effective anticancer agents. In recent years, organometallic compounds have emerged as promising candidates for anticancer therapies. Ferrocifen, a hybrid of ferrocene and phenol, has shown potential against resistant cancers and is currently in preclinical trials; thus, the ferrocene core serves as a valuable framework for developing new anticancer agents. This chapter highlights recent developments in ferrocene-based hybrids as potent anticancer agents and discusses their structure-activity relationships. This information will aid researchers in designing and developing more selective and specific novel anticancer agents to combat drug resistance.

Keywords: Amino acids, Anticancer potential, Artemisinin, Azoles, Carbohydrates, Coumarin, Drug resistance, Ferrocene, Molecular hybridization, Phenol, Pyrimidines, Puinolines, Steroid, Structure activity relationship.

INTRODUCTION

Cancer is the second leading cause of death worldwide, next to cardiovascular diseases, and continues to be a substantial health concern in the 21st century. The World Health Organization (WHO) reported around 20 million new cancer cases and 9.7 million deaths related to cancer worldwide in 2022 [1]. Furthermore, about 53.5 million individuals were living within five years of their cancer diagnosis. It is estimated that one in five people across the globe will face a cancer diagnosis in their lifetime, with one in nine men and one in twelve women

* **Corresponding author Anindra Sharma:** Department of Chemistry, Lalit Narayan Mithila University, Darbhanga, Bihar, India; E-mail: anindrasharma2007@gmail.com

Abhay Nanda Srivastva (Ed.)
All rights reserved-© 2026 Bentham Science Publishers

ultimately succumbing to the disease. The global cancer burden is expected to increase by 77%, with over 35 million new cases anticipated by 2050, compared to 2022 statistics [2]. This rising burden is driven by an aging and growing population, alongside shifts in risk factors due to socioeconomic development. Key contributors include tobacco use, alcohol consumption, obesity, and environmental factors such as water and air pollution. The most prevalent cancers globally affect the breast, lungs, prostate, colorectum, liver, stomach, esophagus, and cervix.

Metal complexes have emerged as promising platforms for drug design, securing an important place in inorganic medicinal chemistry [3]. Cisplatin, the most well-known metal complex, is widely used in cancer treatment, either as a standalone therapy or in combination with other chemotherapeutic agents or radiotherapy [4, 5]. Several cisplatin derivatives, for example, oxaliplatin, carboplatin, heptaplatin, nedaplatin, and lobaplatin, have been developed and are now integral to standard cancer chemotherapy regimens. However, their use remains limited to specific cancer types, and patients often experience significant side effects, with only modest improvements in survival. Furthermore, the development of drug resistance diminishes the effectiveness of these metallodrugs, highlighting the need for more effective and less toxic alternatives in cancer treatment [6, 7]. Non-platinum metal complexes have shown promising results in preclinical and clinical studies, and some have been approved by the WHO for treating various cancers, offering hope for future therapies [8].

Iron, an essential transition metal, is vital to biological systems. It is a key component of the heme complex, which is found in cytochrome proteins that facilitate redox reactions in oxygen-transporting proteins like hemoglobin, leghemoglobin, and myoglobin. Iron's involvement in cancer cell growth makes it an important target for therapeutic development [9, 10]. Its versatile redox chemistry has led to the exploration of iron-based coordination complexes as potential cancer treatments. Bleomycin, an iron-chelating antibiotic, is a notable example, as it induces oxidative DNA damage in the presence of oxygen and H_2O_2, leading to cancer cell death [11, 12]. Similarly, iron(II) chelators have shown significant potential, particularly in overcoming drug resistance, indicating their broader application in cancer therapies [13].

Among iron-based metal complexes, ferrocene is notable for its stable, sandwich-like structure. It exhibits resistance to heat, air, and light, possesses low toxicity, is cost-effective, and demonstrates reversible redox behavior, ligand exchange capabilities, and catalytic activity [14]. These characteristics make ferrocene derivatives versatile scaffolds for drug discovery, with a broad spectrum of pharmacological activities, including antibacterial, antifungal, antimalarial, antitubercular, antiviral, and anticancer effects [14, 15]. Ferrocifen, a hybrid of ferrocene and phenol, shows promise in treating drug-resistant cancers and is currently in preclinical trials [16], underscoring the potential of ferrocene as a framework for novel anticancer agents. Beyond its cytotoxic role, the chemistry of ferrocene also connects directly to drug-delivery applications through both coordination and redox mechanisms. The reversible Fe(II)/Fe(III) redox couple (ferrocene/ferrocenium) can act as a redox trigger: oxidation increases hydrophilicity, leading to the disassembly of hydrophobic carriers or cleavage of redox-sensitive linkers, thereby enabling controlled drug release; reduction reverses this process and stabilizes the carrier system [15, 17, 18]. Coordination chemistry further expands these applications, as ferrocenyl ligands can be incorporated into metal–organic frameworks, supramolecular cages, or polymeric assemblies to create nanostructures with tunable encapsulation and release properties, targeted delivery, and theranostic potential [17]. In addition, ferrocene has been utilized directly as a redox-activated pharmacophore, for example, *N*-alkylaminoferrocene prodrugs are selectively activated in the oxidative microenvironment of tumors, thereby improving therapeutic specificity and even allowing radiolabeling for imaging [18]. Considering this dual role of ferrocene in both therapeutic activity and delivery strategies, extensive efforts have been directed toward designing ferrocene derivatives with enhanced anticancer efficacy. This chapter provides an overview of recent advances in ferrocene-based hybrids, particularly those obtained through molecular hybridization, for the development of effective agents to overcome drug resistance.

MECHANISMS OF ANTICANCER POTENTIAL

The incorporation of ferrocene into bioactive scaffolds has attracted considerable attention due to its unique chemical and biological properties. The Fe(II)/Fe(III) redox couple confers reversible electron-transfer ability, while the sandwich structure provides remarkable chemical stability and lipophilicity [17]. These features allow ferrocene not only to serve as a structural motif but also to actively contribute to cytotoxic mechanisms when embedded in hybrid molecules. Over the past two decades, several mechanistic pathways have been proposed to explain the anticancer activity of ferrocene-based hybrids, among which three have emerged as the most prominent: generation of Reactive Oxygen Species (ROS), DNA interaction coupled with topoisomerase inhibition, and tubulin binding leading to disruption of mitotic progression [19].

Ferrocene-based compounds exhibit potent anticancer activity through these interconnected mechanisms. One primary pathway involves reversible redox cycling of the ferrocene unit within the cellular environment, interacting with endogenous reductants such as glutathione and NADPH to generate ROS, including hydroxyl radicals, hydrogen peroxide, and superoxide anions [20, 21]. These ROS induce cytotoxicity by triggering apoptosis and, depending on the dose, senescence, a permanent halt in cell proliferation mediated by DNA damage and activation of tumor suppressor proteins such as P53. Cancer cells, which maintain elevated basal ROS levels due to altered metabolism, are particularly susceptible to additional oxidative stress, resulting in biomolecular damage, mitochondrial dysfunction, and cell death [22]. Various ferrocene derivatives, including ferrocene–triazole and ferrocene–trioxane conjugates, significantly elevate intracellular ROS, with cytotoxicity mitigated by ROS scavengers, and ROS also modulate the cell cycle, gene expression, hypoxic responses, lipid peroxidation, intracellular calcium signaling, and enzymes such as protein kinase C, histone deacetylases, and DNA methyltransferases [19, 23].

A second mechanism involves direct interaction with DNA and inhibition of DNA-processing enzymes [17]. Ferrocene conjugated to planar aromatic pharmacophores such as quinoline, coumarin, or other heterocycles can intercalate between DNA base pairs, disrupting its structure [24, 25]. The ferrocene moiety enhances cellular uptake and contributes to redox-mediated DNA damage [26]. Several ferrocene–quinoline derivatives stabilize DNA–topoisomerase II cleavage complexes, preventing religation of DNA strands and inducing double-strand breaks [27, 28]. This triggers cell-cycle checkpoint activation, halts proliferation, and promotes apoptosis, similar to anthracycline drugs, but further potentiated by ROS produced from the ferrocene unit [16]. Redox stress can also impair DNA repair pathways and amplify genomic instability, reinforcing cytotoxicity [29, 30].

A third mechanism targets microtubule dynamics by inhibiting tubulin polymerization, which is essential for mitotic spindle formation and chromosome segregation [31]. Ferrocene–chalcone and ferrocene–quinoline derivatives strongly bind to the colchicine site of tubulin, disrupting spindle assembly, causing G2/M phase cell-cycle arrest, and activating apoptotic pathways [16, 32, 33]. In addition to direct tubulin inhibition, ferrocene enhances lipophilicity and cellular uptake while generating ROS that destabilize cytoskeletal proteins, creating a dual mechanism that combines structural microtubule disruption with oxidative stress [17, 26].

These three mechanisms are highly interconnected: ROS exacerbate DNA damage and destabilize microtubules, while DNA/topoisomerase interference and microtubule disruption converge on apoptosis, all potentiated by oxidative imbalance [26, 34, 35]. Thus, ferrocene functions not merely as a structural scaffold but as a multifunctional pharmacophore that amplifies multiple anticancer pathways, explaining the pronounced cytotoxicity of ferrocene-based hybrids and supporting their development as promising candidates for next-generation anticancer agents.

FERROCENE-AMINO ACID HYBRIDS

Amino acids, the fundamental components of proteins, are essential in drug design and medicinal chemistry, as proteins control nearly all cellular functions [36]. The sequence of amino acids governs a protein's structure and function. Drugs linked to amino acids or peptides can target specific receptors on cancer cells, potentially overcoming multidrug resistance in chemotherapy [37]. Thus, ferrocene-amino acid hybrids, which maintain biological activity, offer a promising strategy for developing novel anticancer agents with enhanced efficacy against both drug-sensitive and drug-resistant cancer cells.

Chakravarty *et al.* have assessed the photo-cytotoxic potential of *N*-ferrocenylmethyl-L-amino acid, *viz.* L-Trptophan, L-Tyrosine, and L-Methonine hybrids **1-3** against HeLa and MCF-7 cancer cell lines (Fig. **1**) [38]. Structure-Activity Relationship (SAR) analysis indicated that including a curcumin ligand was essential for strong activity, as substituting it with an acetylacetonate moiety led to a significant decline in activity (IC_{50}: >100 μM). The hybrids **1a**, **2a**, and **3a** exhibited notable photocytotoxic effects when exposed to visible light (400-700 nm), with IC_{50} values ranging between 2.9 and 15.7 μM, demonstrating greater efficacy than cisplatin (IC_{50}: 68.7 μM) against the same cell lines. Mechanistic studies suggested that these hybrids induced apoptosis to trigger cell death. The same research group further demonstrated that 2-(9-anthryl)-1H-imidazo[4,5-f] [1, 10]phenanthroline, 2-(1-pyrenyl)-1H-imidazo[4,5-f] [1, 10]phenanthroline, and dipyrido-[3,2-a:2',3'-c]phenazine can also serve as ligands [39, 40]. Hybrid **4**, when exposed to visible light (λ_{max}: 400-700 nm), exhibited remarkable photo-induced cytotoxicity against MCF-7 and HeLa cancer cell lines, with IC_{50} values of 0.26 μM and 0.7 μM, respectively.

The ferrocene-L-alanine/glycine hybrids **5** (IC_{50}: 2.84–46.5 μM) demonstrated substantial cytotoxic activity against MCF-7 cells, with hybrids **5a**, **5b**, and **5e** (IC_{50}: 2.84–11.1 μM) showing better potency than standard drug Cisplatin (IC_{50}: 16.3 μM) [41, 42]. SAR analysis showed that hybrids with the L-alanine moiety were generally more effective than those with glycine, especially when a pentafluorophenyl group was included. The most active compound, hybrid **5b** (IC_{50}: 2.84 μM), was 5.7 times more potent than Cisplatin and caused oxidative damage *via* a Reactive Oxygen Species (ROS)-mediated pathway.

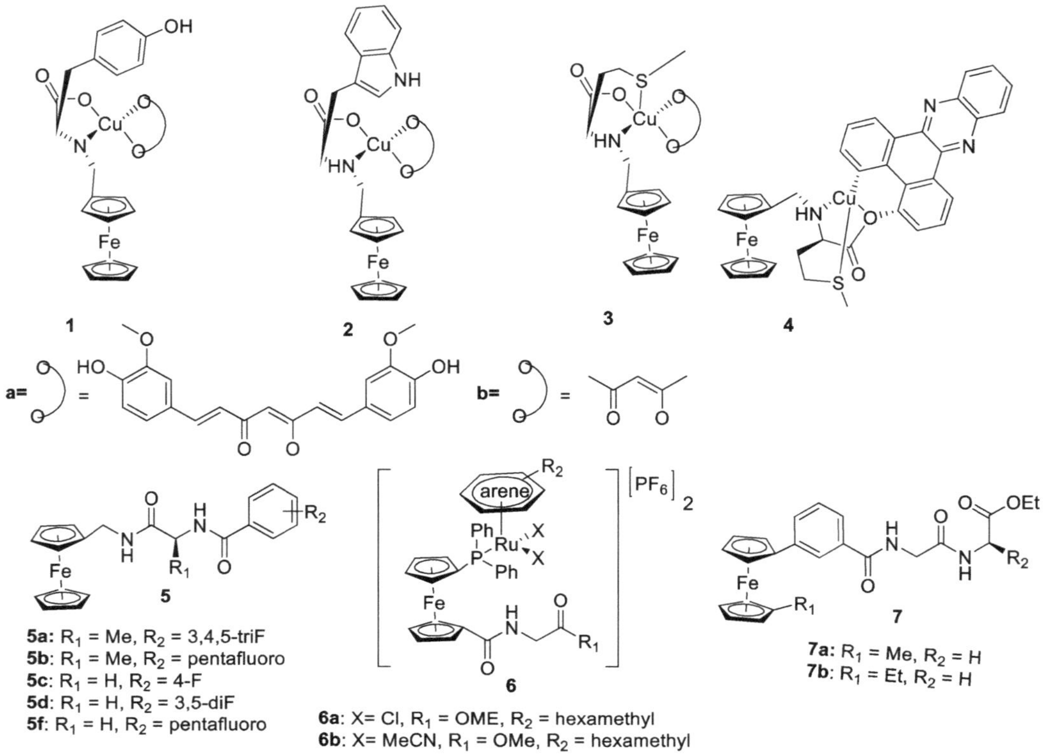

Fig. (1). Ferrocene-amino acids hybrids **1-7**.

Tauchman *et al.* have synthesised a series of phosphinoferrocene-glycine hybrids **6** with notable anticancer activity (IC_{50}: 4.1–35.4 μM) against human ovarian cancer cell lines A2780 and cisplatin-resistant cell lines A2780cisR [43]. SAR analysis indicated that the presence of a methyl ester at the R_1 position significantly enhanced activity compared to the amide group, while a methyl Group at the R_2 position was also favorable for improving potency. In particular, hybrids **6a** and **6b** (IC_{50}: 4.1–8.7 μM) demonstrated the highest effectiveness against both the drug-sensitive A2780 and the drug-resistant A2780cisR cell lines, highlighting their potential as promising candidates for further development in anticancer drug design. Harry *et al.* reported that ferrocenyl dipeptide esters **7** exhibited micromolar anticancer activity (IC50: 2.6-20.1 μM) against the H1299 Non-Small Cell Lung Cancer (NSCLC) cell line. SAR study of these hybrids revealed that a short alkyl side chain at the R_1 position on the ferrocene scaffold enhanced anticancer activity, whereas alkyl and benzyl substitutions at the R_2 position reduced potency compared to the unsubstituted derivatives [44, 45]. Hybrids **7a** and **7b**, with IC_{50} values of 2.6 μM and 6.1 μM, respectively, exhibited comparable activity to Cisplatin (IC_{50}: 1.5 μM), making them promising candidates for further investigation and development.

FERROCENE-COUMARIN HYBRIDS

Coumarins are naturally occurring organic molecules that exhibit various biological activities, including antimicrobial, antibacterial, anti-inflammatory, antioxidant, neuroprotective, anticoagulant, and anticancer properties [46]. Their structural and physicochemical properties make them valuable scaffolds in pharmaceutical and chemical biology. In addition to their pharmaceutical applications, coumarins are used in industries such as food, fragrances, and cosmetics [47]. As a result, molecular hybridization of coumarin with ferrocene has the potential to yield promising lead molecules for the development of novel anticancer agents. Khanye *et al.* reported that the ferrocene-novobiocin hybrid **8** (IC_{50}:1.06-20.7 μM) exhibited substantial anticancer activity against HCC38 breast cancer cells, compared to Paclitaxel (IC_{50}: 0.029 μM) (Fig. **2**) [48]. Among the series, hybrid **8d** was the most effective, with an IC_{50} of 1.06 μM. SAR analysis indicated that the presence of an 8-methyl group was crucial for activity, and in most cases, the compounds displayed approximately 10-fold greater potency than their hydrogen-containing counterparts. Additionally, ferrocenyl analogues demonstrated superior efficacy over non-ferrocenyl variants. Similarly, SAR analysis of ferrocene-novobiocin hybrids **9** (IC_{50}: 11.8-51.4 μM) and **10** (IC_{50}: 13.8-108.9 μM), which feature a piperidine unit in place of the aryl group, also confirmed the importance of the 8-methyl group [49]. The substituent attached to the nitrogen atom followed the anticancer activity trend: H > Boc > Me. Most of these hybrids demonstrated better potency against breast cancer cell lines (MDA-MB-231) compared to Novobiocin (IC_{50}: 205.1 μM), although these hybrids were still much less potent than Paclitaxel (IC_{50}: 0.029 μM).

Ferrocene-coumarin hybrids **11** (IC_{50}: 1.09-62.57 μM) demonstrated broad-spectrum anticancer activity against various cell lines, including MCF-7, SGC-7901, BIU-87, Eca-109, EC-9706, and Jurkat [50]. Despite their effectiveness, most hybrids exhibited lower potency than the reference drug Doxorubicin (IC_{50}: 4.50-8.56 μM). SAR analysis showed that the insertion of a carbonyl group at the X position led to a decrease in activity, while substituting the ester group (Y = O) with an amide group (Y = NH) did not notably improve activity. Among these, hybrid **11a** (IC_{50}: 5.24 and 7.46 μM) exhibited comparable potency to Doxorubicin (IC_{50}: 6.09 and 5.44 μM) against BIU-87 and SGC-7901, identifying it as a potential candidate for further development.

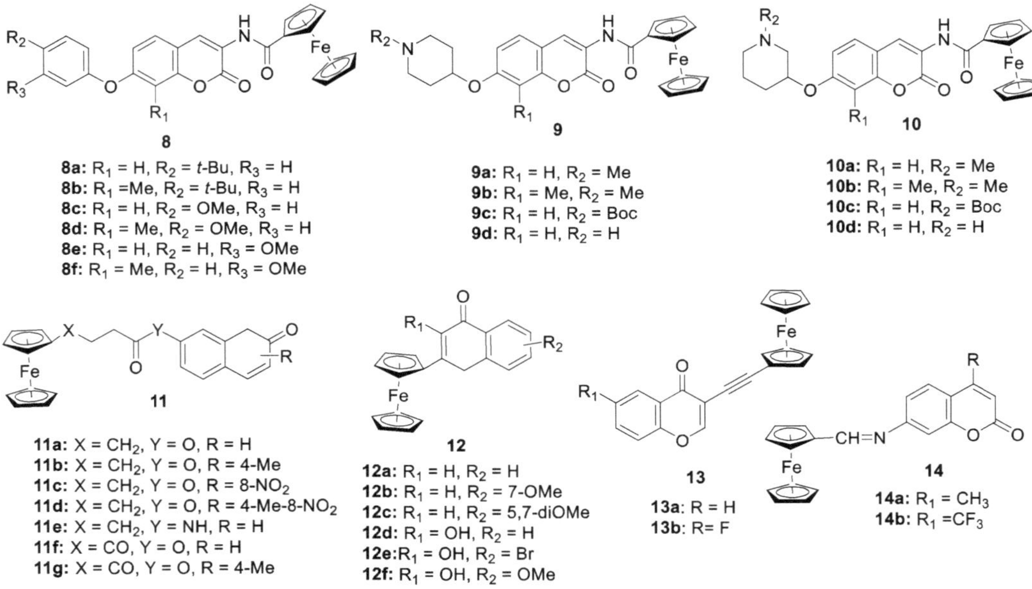

Fig. (2). Ferrocene-coumarin hybrids **8-13**.

Boumendjel *et al.* reported that ferrocene-flavone hybrids **12a-c** (EC_{50}: 6.3 to >100 μM) exhibited notable anticancer potency against wild-type NCI-H69, BHK21, and drug-resistant H69AR and BHK21-MRP1 cell lines [51]. Their efficacy was similar to Verapamil (EC_{50}: 6.7 to >100 μM), which induces collateral sensitivity *via* the MRP1 transporter. SAR analysis showed that adding a -OH group at the C-3 position of the flavonoid unit reduced activity [52]. Notably, these hybrids exhibited better potency against drug-resistant BHK21-MRP1 and H69AR cells (EC_{50}: 6.3-25 μM) compared to wild-type NCI-H69 and BHK21 cells (IC_{50}: 43.8 to >100 μM), highlighting their potential in combating drug-resistant cancers.

Kowalski *et al.* reported that the ferrocene-flavonoid hybrids **13** (IC_{50}: 18.7-43.5 μM) demonstrated potential anticancer activity against MDA-MB-231, MCF-7, RC-124, and HT-29 cell lines [53]. Hybrid **13b** exhibited slightly higher activity than complex **13a** across all tested cancer cell lines, likely due to the presence of a fluorine atom at position 6 of the chromone scaffold. These hybrids demonstrated better activity than Coumestrol (IC50: 48.6 > 100 μM) but were still less potent than Cisplatin (IC_{50}: 1.6-7.7 μM).

Recently, Taner *et al.* investigated the anticancer effects of ferrocene–coumarin conjugates **14a-b** against metastatic colorectal carcinoma (CRC) DLD-1 cells [54]. Both compounds suppressed cell growth in a concentration-dependent fashion, displaying IC_{50} values of 64.3 ± 2.4 and 67.1 ± 3.1 μM, respectively. Importantly, they exhibited negligible toxicity toward normal colon epithelial CCD-18Co cells at equivalent doses ($p < 0.001$), indicating a promising selectivity profile. Fluorescence imaging further revealed that the hybrids accumulated in both the cytoplasm and nucleus of the cancer cells. Overall, these results highlight the potential of conjugates **14a-b** as selective candidates for colorectal cancer treatment.

FERROCENE-STEROID HYBRIDS

The adrenal glands, located above the kidneys, convert hormones into steroids, which act as signalling molecules. Their structural framework is vital for designing target-based anticancer drugs due to their versatility and broad biological activity [55]. Several anticancer agents, such as Galeterone, Aromasin, and Fulvestrant, have been developed using steroidal pharmacophores. Therefore, hybrids of ferrocene and steroids may offer a promising approach to develop new anticancer drug candidates to overcome multidrug resistance.

Meléndez *et al.* reported that ferrocene steroid hybrid **15-18** exhibited varying anti-proliferative effects on the HT-29 and MCF-7 cell lines (Fig. **3**) [56]. 16-ferrocenylidene-3α-hydroxy androstan-17-one **15** was particularly potent, with an IC_{50} of 1.2 ± 0.4 μM, whereas the 3β-isomer **16** exhibited low potency on HT-29 cell lines (IC_{50}: 18 ± 0.3 μM). Hybrid **15**,**16** and **17** (IC_{50}: 14±3, 20±1, and 27 ± 2 μM, respectively) exhibited stronger antiproliferative activity compared to tamoxifen (47 ± 2 μM) against MCF-7 cell lines, whereas hybrid **18** exhibited minimal anti-proliferative activity against HT-29 cell lines and MCF-7 cell lines with IC_{50}: 126 ± 7 μM and (IC_{50}: 124 ± 22 μM, respectively. The cytotoxic potential of these complexes likely arises from a synergistic contribution of receptor recognition and ROS-mediated cytotoxicity.

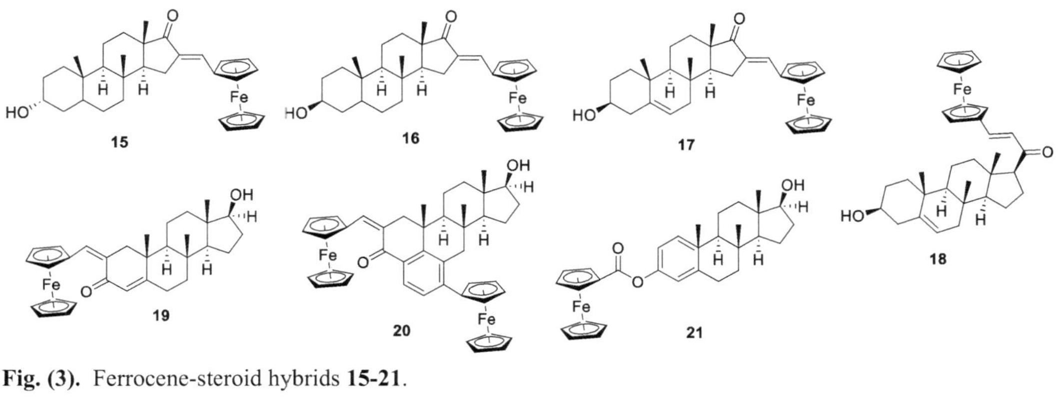

Fig. (3). Ferrocene-steroid hybrids **15-21**.

Manosroi *et al.* reported that the ferrocene-steroid hybrids **19** (GI_{50}: 223 nM) and **20** (GI_{50}: 271 nM), derived from testosterone and methyltestosterone, exhibited dose-dependent antiproliferative activity against HeLa cells, comparable to the reference Doxorubicin (GI_{50}: 250 nM) [57]. Further studies revealed that incorporating an ester linker between the ferrocene and steroid skeleton enhanced activity, with hybrid **21** showing impressive IC_{50} values of 9 nM against MCF-7 and 24.4 nM against HT-29 cancer cell lines [58]. These hybrids show potential for advancing treatments for hormone-related cancers.

FERROCENE-AZOLE HYBRIDS

Azoles and their derivatives or analogs are highly significant and have found wide applications in materials science, agriculture, biology, industry, and medicinal chemistry. These compounds have shown remarkable results in antifungal, antibacterial, antimalarial, antimicrobial, antioxidant, and anticancer activities [59, 60]. Ferrocene-containing azole derivatives, in particular, exhibit notable bioactive properties, including antioxidant, antifungal, antibacterial, antimalarial, and anticancer effects [61]. Thus, combining ferrocene with azole presents a promising approach to developing novel anticancer agents.

Imidazole and its derivatives have displayed significant attention in medicinal chemistry due to their broad range of pharmacological activities, such as antifungal, antibacterial, anti-tubercular, and anticancer properties [62]. Therefore, ferrocene-imidazole hybrids hold potential as novel anticancer agents to combat drug-resistant cancers. Schobert *et al.* reported that the ferrocenyl hybrids of *N*-heterocyclic carbene complexes **22-24** exhibited significant to excellent anticancer activity against seven different cell lines, including MDR cell lines, with IC_{50} values ranging from 0.06 to 19.5 μM (Fig. **4**) [63]. SAR analysis revealed that the *N*-ethyl analog, hybrid **22b**, showed significantly greater potency, with IC_{50} values between 0.2 and 4 μM over 72 hours, compared to the *N*-methyl hybrid **22a**. Furthermore, replacing the chloride ligand with either a triphenylphosphine or another NHC ligand, as seen in the cationic complexes **23** and **24**, was found to enhance anticancer activity, with IC_{50} values falling in the submicromolar range. Among these, complex **24** demonstrated slightly higher potency compared to complex **23**. Chiniforoshan *et al.* reported that Hybrid **25** (IC_{50}: 80–940 μM) demonstrated remarkable efficacy against several cancer cell lines, including HT29, HCT-15, MDA-MB-231, MCF-7, HL-60, human ovarian 2008, and cisplatin-resistant cell lines [64]. Its potency was between 15.3 and 263.3 times higher than that of Cisplatin (IC_{50}: 4.73–90.12 μM) in all tested cell lines. Hybrid **25** was found to effectively inhibit Thioredoxin Reductase (TrxR) and Glutathione Reductase (GR), two key enzymes involved in maintaining cellular redox balance. Uptake studies confirmed that the metal center in **25** is retained with high biological stability, supporting its efficient delivery into cancer cells. In HCT-15 cells, treatment with **25** led to a marked increase in intracellular ROS levels, which in turn triggered apoptotic pathways and resulted in significant cancer cell death. These results indicated that the presence of an Au center in the multinuclear scaffold plays a vital role in improving the anticancer potency of the metallodrugs.

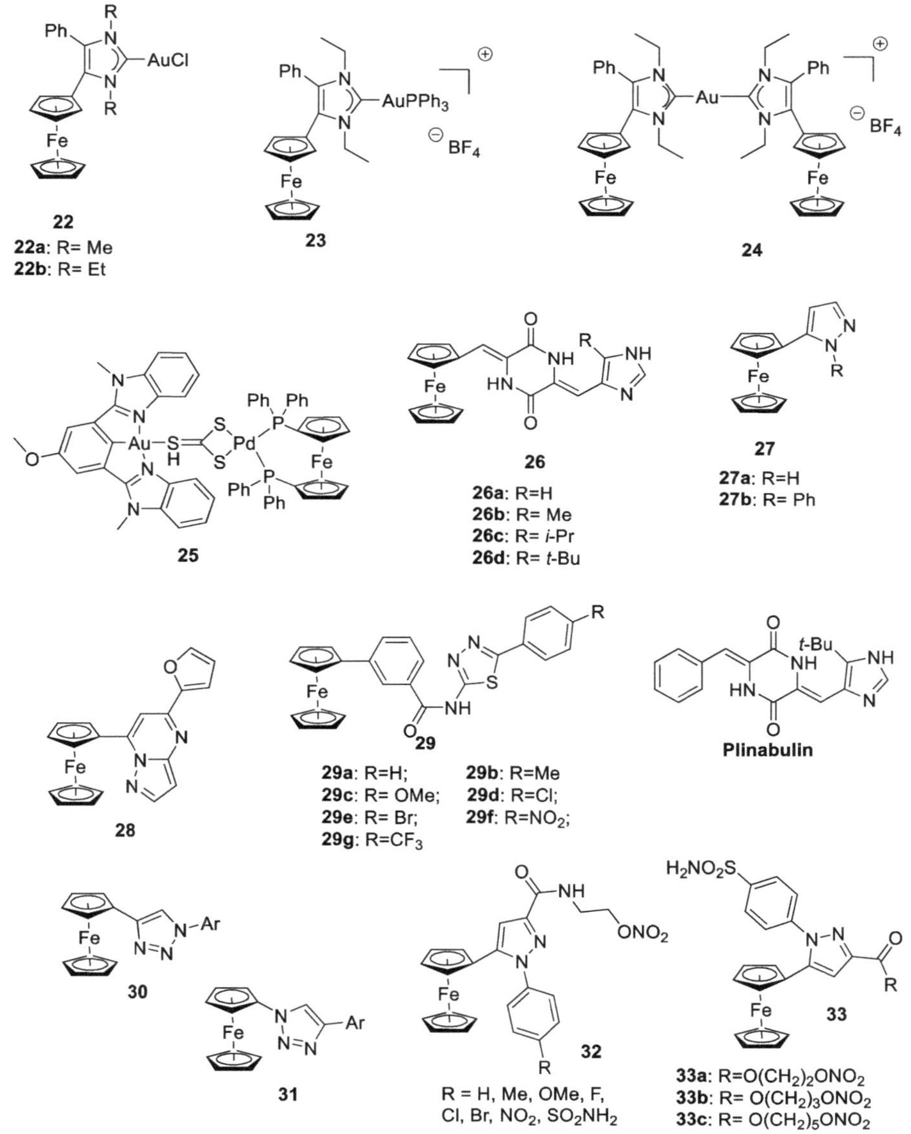

Fig. (4). Ferrocene-azoles hybrids **22-33**.

Plinabulin, a synthetic analogue of 2,5-piperazinedione (phenylahistin), is a promising anticancer drug candidate [65]. It interacts with the colchicine site on tubulin, disrupting tumor blood vessels and obstructing the blood supply to cancer cells, which ultimately results in cell death. Plażuk *et al.* evaluated various ferrocenyl analogues of plinabulin **26** for anticancer activity against MCF7, HepG2, and SW620 cell lines, as well as drug-resistant SW620 variants: SW620C (cisplatin), SW620D (doxorubicin), SW620E (etoposide), SW620M (methotrexate), and SW620V (vincristine) [66]. It was observed that hybrids **26c** and **26d**, with bulky alkyl groups on the imidazole moiety, exhibited similar effectiveness across all tested cell lines. However, their toxicity (IC_{50} in the micromolar range) differed, being highest in SW620 cells and lowest in HepG2 cells. SAR analysis confirmed that the size of the alkyl group R strongly influenced cytotoxicity, following the order t-Bu > i-Pr > Me > H. Furthermore, the presence of both ferrocene and imidazole units was crucial for maintaining the anticancer activity of the synthesized compounds.

Zora *et al.* reported that ferrocene-pyrazole hybrid **27a** (IC_{50}: 27.03 and 34.51 μM) demonstrated greater anticancer activity than the hybrid **27b** (IC_{50}: 42.24 and 85.51 μM) against MCF-7 and MDA-MB-231 cell lines, respectively, indicating that the inclusion of a phenyl group in the pyrazole fragment reduced its efficacy [67]. Studies revealed that these hybrids can reduce cell viability and trigger either apoptosis or necrosis in breast cancer cells, while exhibiting minimal toxicity to healthy breast epithelial cells. Furthermore, these hybrids suppress breast cancer cell viability by inhibiting the PI3K/Akt and ERK1/2 signalling pathway. Recently, Shard *et al.* reported that ferrocene-derived pyrazolopyrimidine **28** (IC_{50}: 7.49 μM) exhibited strong antiproliferative activity, particularly against cisplatin-resistant cancer cell lines [68]. Moreover, this hybrid activated tumor Pyruvate Kinase M2 (PKM2), an essential enzyme in cancer metabolism, and induced apoptosis, resulting in cell cycle arrest during the S phase.

Xu *et al.* assessed a series of *m*-ferrocenylbenzoylthiadiazole hybrids **29** for their anticancer activity across four cancer cell lines: EC-9706, Eca-109, SGC-7901, and Jurkat [69]. SAR analysis revealed that compounds with electron-donating groups like OMe or Me exhibited no significant activity. In contrast, introducing electron-withdrawing groups (*e.g.*, Cl, Br, NO_2, or CF_3) led to notable inhibitory effects, particularly against EC-9706 and Eca-109 cell lines. Compound **29f**, containing a NO_2 group, exhibited the highest potency against EC-9706 ($IC_{50:}$ 18.61 μM). Furthermore, compound **29e** (IC_{50}: 28.17 μM) demonstrated potent activity against gastric cancer cells (SGC-7901), while compound **29g** displayed the highest efficacy against Eca-109 cells (IC_{50}: 15.51 μM). However, their efficacy was lower compared to the reference drug Doxorubicin, which had IC_{50} values of 8.56 and 6.52 μM, respectively. The study emphasized that stronger electron-withdrawing substituents enhanced anticancer activity, highlighting their relevance against the tested cancer cell lines.

The triazole framework enhances the pharmacokinetic, pharmacological, and physicochemical properties of drug-like molecules, as evidenced by triazole-based drugs like Cefatrizine and Carboxyamidotriazole, which show significant anticancer potential [70, 71]. Therefore, hybridizing ferrocene with triazole units could offer promising therapeutic options for treating drug-resistant cancers. Plażuk *et al.* screened various ferrocene-*C*-triazoles **30** and ferrocene-*N*-triazoles **31** for their anticancer activity against MCF-7 and HCC38 breast cancer cell lines [72]. These hybrids exhibited moderate anticancer activity against the HCC38 cell line (IC_{50}:15.3-48.9 μM) and lower potency against the MCF-7 cell line (IC_{50}:84-98 μM). SAR analysis revealed that the ferrocenyl-*C*4-triazolyl derivative exhibited better anticancer activity against HCC38 compared to the ferrocenyl-*N*1-triazolyl derivatives.

Cyclooxygenase (COX), a key enzyme in prostaglandin biosynthesis, impacts inflammation, gastrointestinal protection, and tumor growth. COX-1 regulates normal physiological functions, while COX-2 is induced by inflammatory stimuli and is overexpressed in many cancers, promoting tumor progression and inhibiting apoptosis [73]. Ferrocene-pyrazole hybrids **32** and **33** demonstrated higher activity against COX-2 (IC_{50}: 1.25-17.09 μM for **32** and 0.28-2.03 μM for **33,** respectively) compared to COX-1 (IC_{50}: 29.09-41.05 μM for **32** and 21.07-47.85 μM for **33,** respectively), positioning them as promising COX-2 inhibitors [74]. These compounds also demonstrated significant anticancer activity against F10, A549, HeLa, and MCF-7 cancer cell lines, with efficacy similar to or better than the reference drug Celecoxib (IC_{50}: 7.55-25.87 μM). SAR analysis revealed that adding a sulfonamide group to the phenyl ring at the *N*-1 site of the pyrazole core improved activity, making hybrids **33** more potent than **32**. In hybrids **33**, ester groups at the R position showed greater effectiveness than amides, and a linear side chain proved more advantageous than a branched one for both ester and amide derivatives. Notably, hybrids **33a-c** with IC_{50}: 0.34-5.78 μM exhibited the highest potency across all tested cancer cell lines while displaying low cytotoxicity towards normal 239t cells (IC_{50}: 171.70-280.42 μM). Mechanistic studies indicated that hybrid **33b** induced apoptosis in HeLa cells *via* mitochondrial depolarization. The antiproliferative effect of **33b** was linked to increased intracellular NO release in HeLa cell lines. *In vivo* studies using a HeLa cell xenograft model in Balb/C nude mice showed that hybrid **33b** (20 mg/kg, intraperitoneally) significantly reduced tumor weight compared to the vehicle control group. Interestingly, the tumor weight in the Celecoxib-treated group (10 mg/kg) was four times greater than that in the hybrid 31b-treated group. Furthermore, mice treated with hybrid **33b** showed no significant changes in body weight, suggesting low toxicity. The potent anticancer activity of hybrid **33b**, both *in vitro* and *in vivo*, along with its low toxicity, suggests it is a strong candidate for further investigation.

FERROCENE-CARBOHYDRATE HYBRIDS

Carbohydrate moieties offer low cost, ready availability, non-toxicity, and easy modification into various functional groups, making them ideal for developing novel synthetic pathways for transition metal complexes containing glycoconjugates [75]. Ferrocene-carbohydrate complexes have attracted significant attention due to their applications as electrochemical probes and potential biological activities [76]. Moreover, the hydrophilicity of these conjugates can be finely tuned through the selective protection or deprotection of hydroxyl groups on the carbohydrate unit. Consequently, ferrocene-carbohydrates open new avenues for developing novel anticancer agents to overcome resistance challenges. Trivedi *et al.* screened three different series of ferrocene-sugar hybrids: one with an amide linker **34**, one with a triazole linker **35**, and a third with an amide-triazole linker **36** against Neuro2a, A549, HeLa, MCF-7, and MDA-MB-231cell lines (Fig. **5**) [77 - 79]. Hybrids **34–36** demonstrated moderate to significant anticancer activity, with IC_{50} values varying between 5.3–11.59, 5.5–39.74, and 2.8–11.31 μM, respectively. Hybrids **36b** (IC_{50}: 2.82 and 11.31 μM) and **36c** (IC_{50}: 7.31 and 3.18 μM) showed promising anticancer potency against MDA-MB-231 and MCF-7 cell lines. SAR studies revealed that, regardless of the linker used, ferrocene-carbohydrate conjugates with a ribofuranoside moiety exhibited similar IC_{50} values for the MDA-MB-231 cell line. Unfortunately, none of the tested ferrocene-sugar hybrids exhibited better activity compared to the reference compounds Cisplatin (IC_{50}: ≤1 μM) and Doxorubicin (IC_{50}: <1–1.2 μM). Replacing the amide bond linking ferrocene and 1,2,3-triazole with ether or thioether (X = O or S, **37a,b** and **38a,b**, IC_{50}: 9.0->200 μM) negatively impacted anticancer activity. In contrast, introducing a selenoether group (X = Se, **37c** and **38c**, IC_{50}: 2.9 and 3.71 μM) enhanced activity against A549 cells. Further investigation revealed that introducing an additional sugar moiety did not enhance anticancer activity. Hybrids **39a,b** (IC_{50}: 38.3 and >100 μM) showed lower efficacy compared to Cisplatin (IC_{50}: 12.9 μM) against A2780 cells.

The *N*-Alkylaminoferrocene (NAAF)-based prodrugs are activated by elevated ROS levels, which are typically found in cancer cells. This activation results in the production of NAAF and p-quinone methide, which work together to elevate oxidative stress and induce cancer cell death. Mokhir *et al.* reported that NAAF-based prodrug, Ferrocene-6-deoxy-6-fluoroglucopyranosyl **40** (IC_{50}: 20-30 μM) demonstrated significant activity against cancer cell lines MCF-7, HL-60, PC-3,

DU-145, and AR42J, making it a promising candidate for further exploration [80]. Radiolabelling with the ^{18}F isotope of fluorine confirmed that boronic acid derivatives play a key role in amplifying ROS, enhancing the anticancer activity of N-alkylamino ferrocene-based prodrugs.

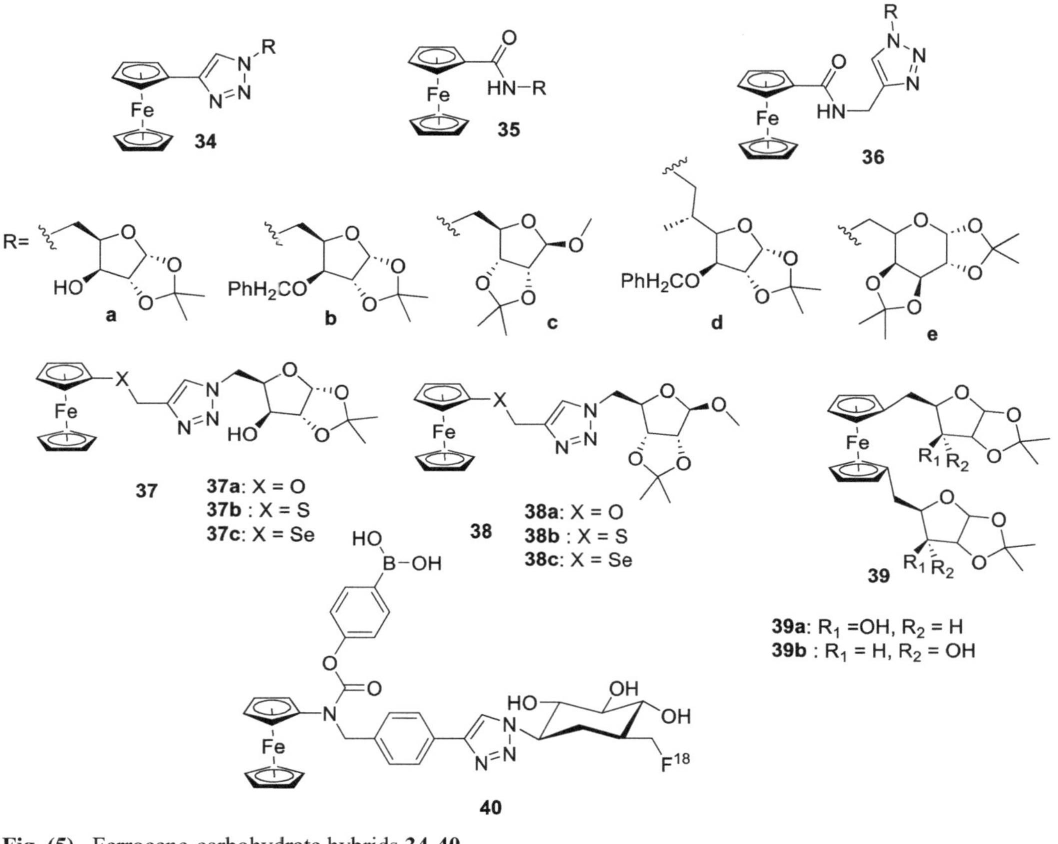

Fig. (5). Ferrocene-carbohydrate hybrids **34-40**.

FERROCENE-PYRIMIDINE HYBRIDS

Pyrimidine bases, present in thymine, cytosine, and uracil, are essential components of DNA and RNA, playing a key role in their wide therapeutic applications. Pyrimidines and their derivatives exhibit diverse biological activities, including antiviral, anti-inflammatory, anti-HIV, antifungal, antihypertensive, analgesic, antioxidant, and anticancer properties. Additionally, some function as CNS depressants and calcium channel blockers [81]. Therefore, combining this pharmacophoric unit with ferrocene could offer promising opportunities for developing new anticancer agents.

Kowalski *et al.* assessed the cytotoxicity of ferrocene-nucleobase conjugates **41** and **42** against MDA-MB-231, HT-29, and non-cancerous L929 mouse fibroblasts using the crystal violet assay (Fig. **6)** [82]. Hybrids **41a** and **42b** exhibited maximum potency, with IC_{50} values ranging from 16.1 to 43.6 μM across these cell lines, comparable to cisplatin (IC_{50}: 2.4–16.1 μM). SAR analysis revealed that only the dinuclear derivative exhibited significant cytotoxicity towards the evaluated cancer cell lines. The data suggest that the incorporation of a sterically hindered, lipophilic Fc-CH_2-Fc group enhances cytotoxic activity, whereas complexes with a single ferrocenyl unit exhibited weak or no activity. The same group further investigated the anticancer properties of ferrocenylthymine-3,6-dihydro-2H-thiopyrane **43a** and ferrocenyl-4-thiothymine-3,6-dihydro- 2H-thiopyrane **43b** against various cancer cell lines, including HT-29, MCF-7, MDA-MB-231, HL-60, and monocytic MonoMac6 cell lines [83]. Hybrid **43b** showed a cytostatic effect on MonoMac6, MCF-7, and HL-60 (IC_{50}: 28.6, 11.1, 88.33 μM, respectively) cells but was inactive against HT-29 and MDA-MB-231. In contrast, compound **43a** was effective across all tested cell lines (IC_{50}: 23.2, 5.1, 8.3, 4.5, and 17.4, 88.33 μM against HT-29, MCF-7, MDA-MB-231, HL-60, and MonoMac6, respectively). The pro-apoptotic activity of both compounds was confirmed using the acridine orange/ethidium bromide double-staining assay. It is suggested that the sulfur atom in hybrid **43b** may hinder the complex's interaction with target proteins, whereas the "natural" thymine in compound **43a** may promote better binding and inhibition.

Morley *et al.* designed and synthesized ferrocene-containing pyrimidine derivatives **44** and **45**, which were screened against three breast cancer cell lines: Triple-Negative Breast Cancer (TNBC) cell lines (BT-549 and MDA-MB-231) and the HER2+ breast cancer cell line (SK-BR-3) [84]. Hybrid **44** showed IC_{50} values of 2.83 μM, 6.67 μM, and 114.30 μM against MDA-MB-231, BT-549, and SK-BR-3 cell lines, respectively, while hybrid **45** exhibited IC_{50} values of 0.55 μM, 1.73 μM, and 10.63 μM against the same cell lines. SAR analysis suggested that the phenyl spacer between the ferrocene and nucleobase enhanced the anticancer activity. Notably, MDA-MB-231 TNBC cells displayed higher sensitivity to both hybrids, with the particularly strong response to hybrid **45** prompting its selection for further investigation.

Simenel *et al.* synthesized ferrocenyl hybrids of thymine, adenine, cytosine, and iodo-cytosine [85]. Among these, the antitumor potency of 1-*N*-ferrocenylmethyl thymine **46** was evaluated *in vivo* using animal tumor models, including carcinoma 755 (Ca755), melanoma B16, and Lewis lung carcinoma. Hybrid **46** exhibited a notable antitumor effect, achieving 70% inhibition of tumor growth in the Ca755 model compared to the control. Additionally, the researchers proposed a synergistic effect by combining hybrid **46** with the established anticancer drug cyclophosphamide for enhanced efficacy against Ca755.

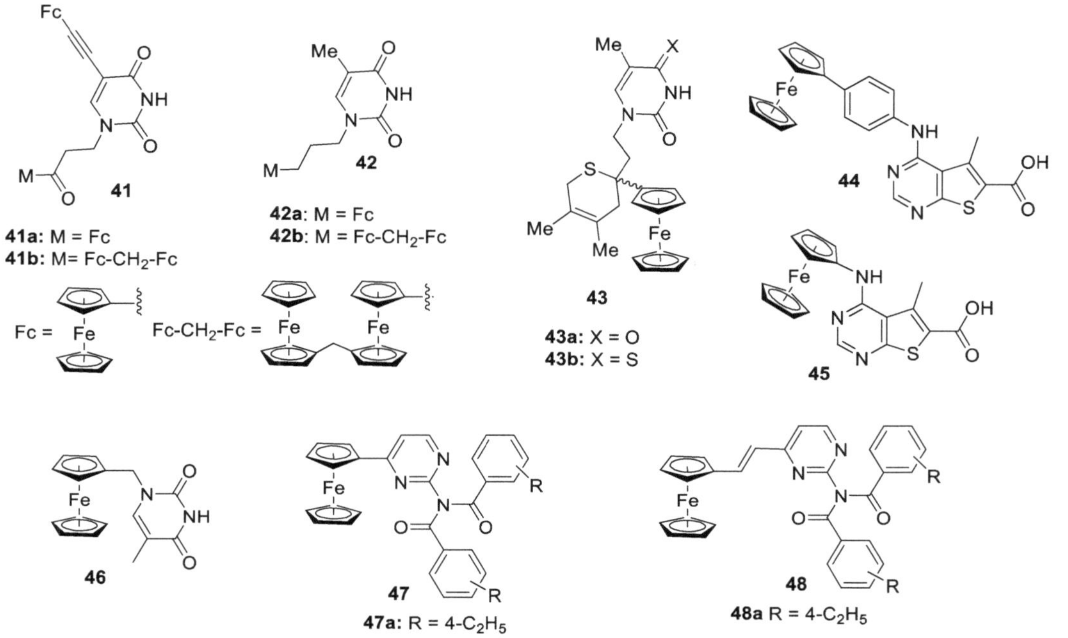

Fig. (6). Ferrocene-pyrimidine hybrids **41-48**.

Ruan *et al.* assessed the *in vitro* antitumor potential of two new series of ferrocene-pyrimidinyl bis-amide derivatives **47** and **48** against B16-F10 and A549 cell lines using the MTT assay, with Celecoxib as a reference compound [86]. The majority of the tested compounds exhibited notable antitumor activity, with IC_{50} values varying between 1.2 μM and 50 μM. SAR analysis revealed that incorporating electron-donating groups into the phenyl ring enhanced activity, and compounds with para-substituted phenyl groups exhibited stronger potency than their ortho-substituted counterparts. Among the compounds, hybrids **47a** (IC_{50}: 1.9 and 16.5 μM) and **48a** (IC_{50}: 18.0 and 1.2 μM) exhibited maximum potency against B16-F10 and A549 cell lines, with potency comparable to or even better than Celecoxib (IC_{50}: 15.22 and 16.03 μM). Further mechanistic studies indicated that hybrid **47a** could induce apoptosis in A549 cells, highlighting its potential as a promising anticancer candidate.

FERROCENE-QUINOLINE HYBRIDS

Quinolines have gained significant attention in medicinal and synthetic chemistry due to their broad utility in developing potent therapeutic agents, making them a crucial structural framework for drug design. They demonstrate a broad spectrum of biological activities, such as antitubercular, anticonvulsant, anticancer, anti-inflammatory, antibacterial, antimalarial, and cardiovascular effects [87]. Drugs such as Neratinib, Amsacrine, Irinotecan, Cabozantinib, Fortinib, and AMG-208 are examples of quinoline-based compounds used to treat various cancers, including breast cancer, acute leukemia, blood cancer, colorectal cancer, thyroid cancer, and prostate cancer [88]. Quinoline and its analogues are known to inhibit key cellular processes such as tyrosine kinases, proteasome activity, tubulin polymerization, topoisomerase, and DNA repair. DNA topoisomerases are well-established targets in cancer therapy, as inhibitors, often referred to as "poisons," form stable complexes with DNA, leading to persistent damage that triggers a series of events resulting in cell death, either through apoptosis or other mechanisms [89]. Quinoline derivatives like Cabozantinib, Voreloxin, and Quarfloxin exert their anticancer effects by inhibiting topoisomerases. Thus, hybridizing ferrocene with quinoline may offer a valuable therapeutic approach for cancer treatment.

Smith *et al.* reported that the ferrocene-quinoline monomeric hybrids **49** and **50** exhibited excellent *in vitro* anticancer activity against WHCO1 esophageal cancer cells, with IC_{50} values of 0.31 μM and 0.74 μM, respectively, compared to cisplatin (IC_{50}: 13.1 μM) (Fig. 7) [90]. SAR analysis indicated that extended conjugation enhanced the anticancer activity. Hybrid **49** exhibited potency against the TK10, UACC62, and MCF-7 cancer cell lines, with GI_{50} values of 1.5, 2.6, and 0.8 μM, respectively, comparable to Etoposide (GI_{50}: 0.8–5.9 μM) [91]. Raić-Malić *et al.* investigated the anticancer activity of *N*-alkylated 4-quinolone **51** and *O*-alkylated quinoline **52** derivatives linked to ferrocene against various cancer cell lines (HeLa, K562, CaCo-2, Raji) and a canine normal cell line (MDCK1), using the MTT assay [92]. Compounds **51a–c** displayed the strongest antiproliferative effects on K562 cells, with IC_{50} values varying between 7.73 and 16.95 μM. Especially, the 6-non-substituted quinolone moiety in **51a** showed greater cytostatic activity across all tested tumor cell lines compared to the methyl-substituted **51b**. Additionally, the methylene linker in **51a** and **51b** improved cytostatic effects relative to the ethylene linker in **51c** and **51d**. Among the 4-*O*-alkylated quinoline-ferrocene derivatives **52a–d** (IC_{50}: 7.9-99.81 μM), compound **52c**, with a chlorine substitution at C-6 of the quinoline, exhibited selective potency against Raji cells (IC_{50}: 7.9 μM) without toxicity to normal MDCK1 cells.

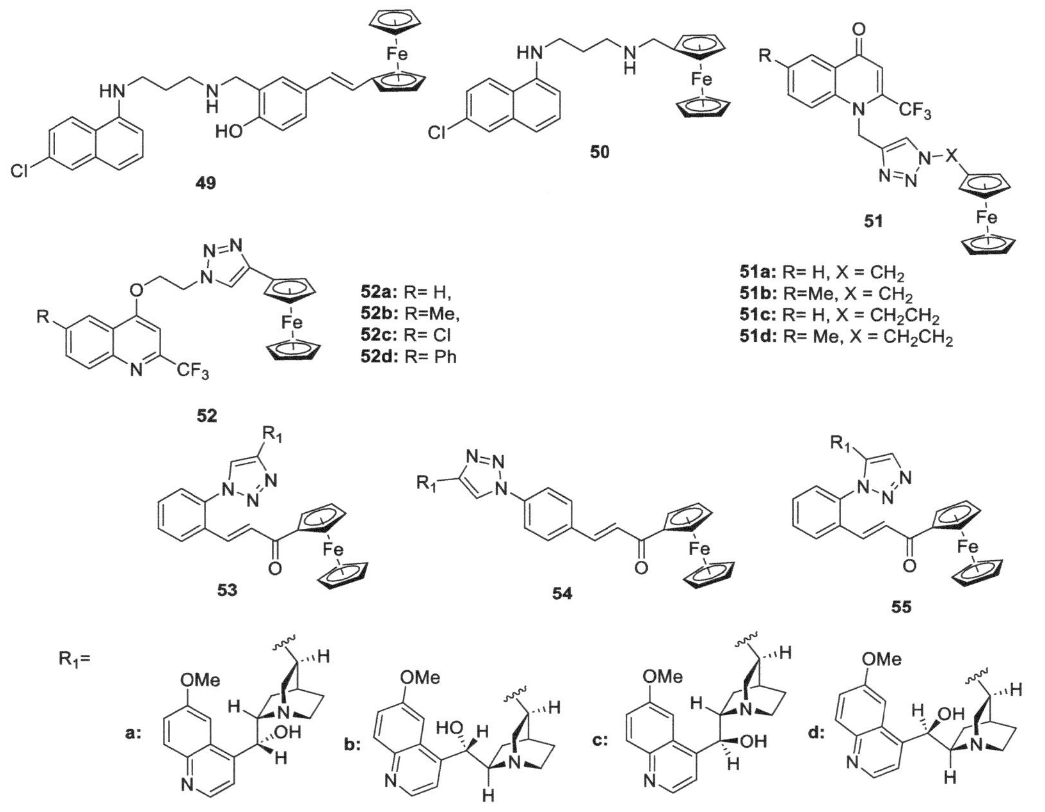

Fig. (7). Ferrocene-quinoline hybrids **49-55**.

Csámpai *et al.* evaluated different series of cinchona-ferrocene hybrids **53-55** for their cytostatic activity against HepG2 hepatoma and HT-29 colorectal adenocarcinoma cells [93]. These compounds showed improved anticancer activity compared to tamoxifen (IC_{50}: 20 μM), with IC_{50} values 0.7-14.3 μM, regardless of the configuration at positions 8 and 9 of the cinchona skeleton or the phenyl ring substitution. Hybrid **55d**, containing the epiquinidine fragment, exhibited the highest potency, with IC_{50} values of 0.7 ± 0.1 μM for HepG2 and 1.5 ± 0.2 μM for HT-29. Importantly, removing the chalcone moiety or replacing it with a CH2 group significantly reduced the activity, emphasizing the crucial role of the chalcone unit in the hybrid's bioactivity. Further evaluation of hybrids **53a-d** against three MDR cancer cell lines and their sensitive counterparts (NCI-H460/R, DLD1-TxR, and U87-TxR) revealed that the MDR phenotype did not affect their activity [94]. This indicates that the ferrocene-quinine and quinidine epimers are likely not recognized as substrates by P-glycoprotein, a key contributor to Multidrug Resistance (MDR) in these cell lines. Hybrids **53b** and

53d were found to elevate ROS levels and cause mitochondrial dysfunction in MDR cancer cells. Cell death assays indicated that these compounds were more effective against MDR cells, showing collateral sensitivity and functioning as potential inhibitors of autophagy. Furthermore, the combination of these compounds with paclitaxel (PTX) enhanced the sensitivity of Multidrug-Resistant (MDR) cancer cells to PTX. These results suggest that ferrocene-quinidine epimers not only selectively target MDR cancer cells but may also help overcome PTX resistance.

FERROCENE-ARTEMISININ HYBRIDS

Artemisinin, a natural 1,2,4-trioxane sesquiterpene derived from *Artemisia annua L.*, is well-known for its antimalarial, anticancer, and antiviral properties. Although its exact mechanism of action remains unclear, the endoperoxide group plays a critical role in cancer and malaria [95]. It reacts with intracellular Fe(II), generating ROS and peroxyl radicals that induce oxidative stress, DNA damage, protein alkylation, and apoptosis [96]. To address potential resistance, hybrid molecules combining artemisinin with other pharmacophoric units have been developed, showing enhanced activity against cancer [97]. In this section, we will explore the anticancer potential of various hybrids formed through molecular hybridization of artemisinin derivatives with ferrocene units.

Reiter *et al.* assessed the cytotoxic effects of hybrids **56-57** on both wild-type CCRF-CEM and multidrug-resistant CEM/ADR5000 leukemia cells that overexpress P-glycoprotein (Fig. **8**) [98]. All five hybrids showed greater cytotoxicity against CCRF-CEM cells (IC_{50} < 0.25 μM) than artemisinin and ferrocene monocarboxylic acid, which did not demonstrate cytotoxic effects within the tested concentration range. This highlights the effectiveness of the hybridization strategy. SAR analysis revealed that ferrocene-artemisinin hybrids with two trioxane units (**56c** and **56d)** were significantly more potent, with IC_{50} values of 0.07 μM and 0.08 μM, respectively, compared to hybrids with one unit (**56a** and **56b**), which had IC_{50} values of 0.13 μM and 0.25 μM, respectively. Hybrid **57** was the most potent compound in the series, exhibiting an IC_{50} of 0.01 μM. The CEM/ADR5000 cells, which display a 2585-fold resistance to doxorubicin (IC_{50} of 23.27 μM compared to 0.009 μM in CCRF-CEM), present a significant challenge in overcoming multidrug resistance. However, hybrids **56-57**, while less effective against CEM/ADR5000 compared to CCRF-CEM, showed 3- to 50-fold higher cytotoxicity than artemisinin. Hybrid **56a** (IC_{50}: 0.53 μM) was the most potent against CEM/ADR5000 cells. Furthermore, the incorporation of an egonol moiety into the ferrocene framework **58**, with an IC_{50} of 0.07 μM, enhanced the potency against wild-type CCRF-CEM cells [99].

However, it significantly reduced efficacy against MDR-CEM/ADR5000 leukemia cells, with an IC_{50} of 0.57 μM.

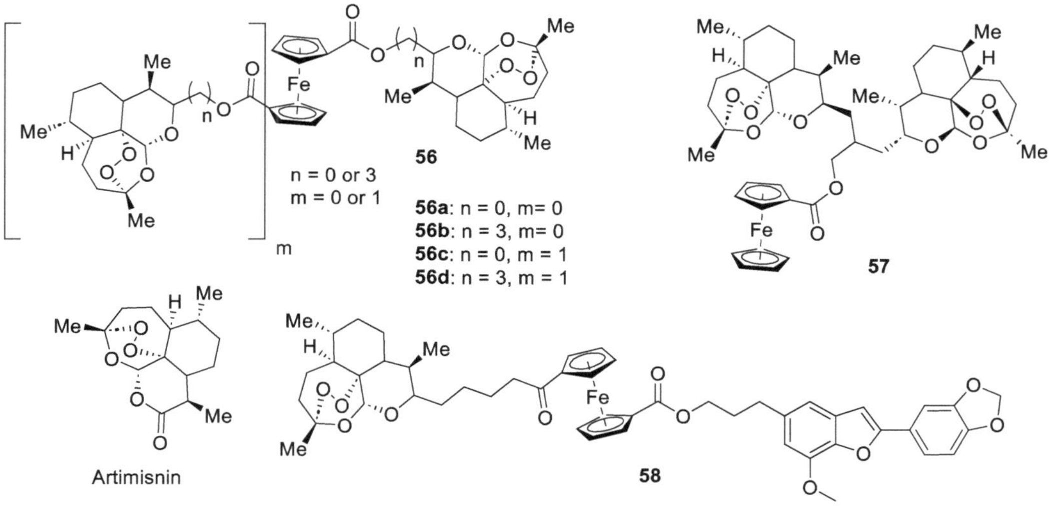

Fig. (8). Ferrocene-artemisnin hybrids **56-58**.

FERROCENE-PHENOL HYBRIDS

Ferrocenyl phenols, particularly ferrocifens **59** developed by Jaouen et al., have demonstrated strong anticancer activity, showing promise in breast cancer treatment (Fig. **9**) [100 - 103]. Electrochemical and chemical oxidation studies suggest a mechanism involving ferrocenyl quinone methides. This section will highlight recent examples of ferrocene-phenol hybrids as potent anticancer agents. Jaouen *et al.* reported that ferrocenyl diphenol complexes **60a-c** showed reasonable cytotoxicity against hormone-independent prostate PC3 (IC_{50}: 2.5–3.2 μM) and breast cancer MDA-MB231 cell lines (IC_{50}: 2.8–4.1 μM), with the effect being stronger in PC3 cells, particularly for the ortho and para diphenol complexes, with IC_{50} values of 2.5 μM and 3.1 μM, respectively [104]. Moreover, the impact of 1 μM of hybrids **60a-c** on the growth of hormone-dependent MCF7 breast cancer cell lines was examined both with and without phenol red. The results showed that the hybrids exerted both cytotoxic (growth-inhibiting) and estrogenic (growth-promoting) effects. Electrochemical analysis revealed that the cytotoxicity of these complexes was associated with the oxidation potential of the ferrocene unit. Lopez *et al.* demonstrated that hybrids **61a–c** exhibited cytotoxicity across multiple cancer cell lines [105]. In A2780 ovarian cancer cells, derivatives **61a–c** showed strong activity, with IC_{50} values of 2.68, 1.66, and 1.86 μM, respectively. In A549 lung cancer cells, **61a** and **61b** also exhibited moderate cytotoxicity, with IC_{50} values of 2.77 and 1.36 μM, respectively, while **61c** was less active, with an IC_{50} of 5.96 μM. The presence of a hydroxyl group appeared crucial, as the methoxy derivative **61a-Me** displayed minimal toxicity (IC_{50} > 10 μM) in both cell lines. The same research group further demonstrated that hybrid **62a** (IC_{50}: 1.07 μM) exhibited stronger toxicity against A2780 ovarian cancer cells compared to its ortho-isomer **61a** [106]. Similarly, **62b** (IC_{50}: 2.23 μM) showed cytotoxicity slightly lower than its ortho analogue **61c** (IC_{50}: 1.86 μM). Both compounds were further evaluated against A549 lung cancer cells, showing moderate potency with IC_{50} values of 3.55 μM for **62a** and 4.89 μM for **62b**, comparable to their ortho-isomer counterparts, **61a** and **61c**.

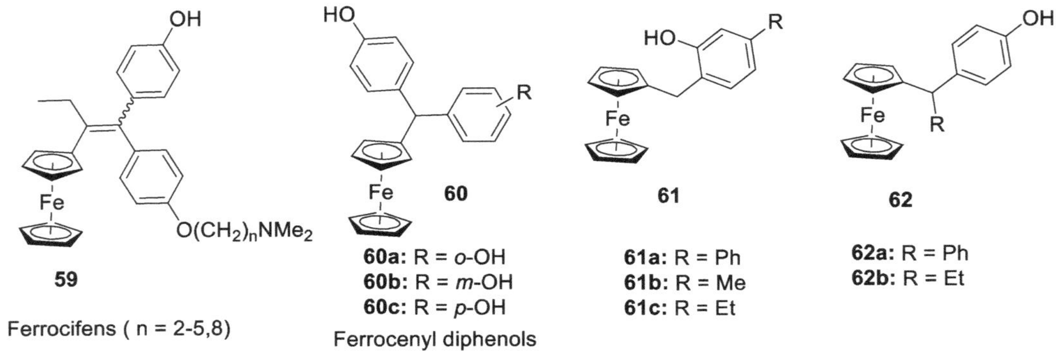

Fig. (9). Ferrocene-phenol hybrids **59-62**..

MISCELLANEOUS FERROCENE HYBRIDS

Several other hybrids of ferrocene with other pharmacophoric active units have reported to exhibit anticancer activity. Plazuk *et al.* evaluated the cytotoxicity of two series of ferrocene-paclitaxel hybrids against various human cancer cell lines (Fig. **10**) [107]. The cytotoxic activity was strongly influenced by the position of the ferrocenyl group, with *n*-benzoyl-substituted derivatives **63a-d** (IC_{50}: 0.009-1.118 μM against SW620 cells) showing the highest potency. Substituting the 3'-phenyl group in paclitaxel's phenylisoserine side chain with a ferrocenyl moiety in derivatives **64a-d** (IC_{50}: 0.490–2.441 μM against SW620 cells) significantly reduced cytotoxicity, lowering activity by approximately two orders of magnitude compared to both paclitaxel (IC_{50}: 0.0151 μM against SW620 cells) and derivative **63a**. Mechanistic studies revealed that compounds inducing strong microtubule polymerization, as confirmed by molecular docking, exhibited greater cytotoxicity, while those with stronger pro-oxidative effects had lower antiproliferative activity. The results suggest that there is no single mechanism of action, as some highly cytotoxic compounds showed weak tubulin binding. Furthermore, 3'-phenyl substitution with a ferrocenyl group, as in **64b** and **64c**, produced low-toxicity compounds capable of overcoming multidrug resistance by avoiding ABCB1 recognition. These findings suggest that ferrocenyl-modified taxanes hold promise as candidates for overcoming drug-resistant cancers.

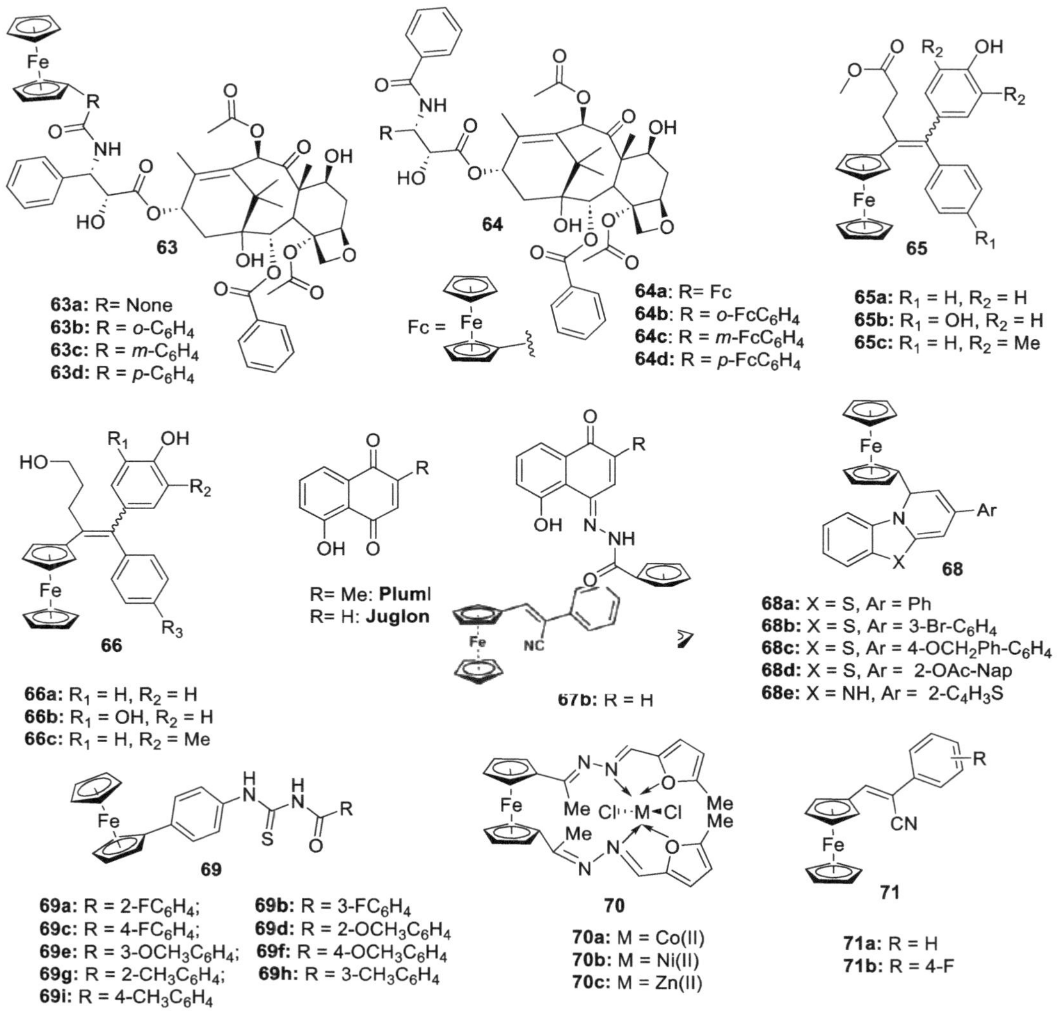

Fig. (10). Miscellaneous ferrocene hybrids **63-71**.

Wang *et al.* evaluated a series of ferrocifen hybrids, ester **65** and alcohol derivatives **66**, against ER-negative MDA-MB-231 cell lines [108]. These hybrids exhibited moderate to excellent anticancer potency, with IC_{50} values varying between 0.11 and 0.49 μM. Hybrid **66b** proved to be the most potent, with an IC_{50} of 0.11 μM, significantly outperforming its ester precursor **65b**, a related diphenol compound, and tamoxifen (IC_{50}: 40 μM), highlighting the critical role of the ferrocenyl group in enhancing antiproliferative effects. Furthermore, **66b** suppressed pancreatic carcinoma (Mia-PaCa) growth with an IC_{50} of 1.23 μM and exhibited strong activity against liver cancer (HepG2) with an IC_{50} of 0.07 μM. This promising activity led to further investigation of its structure to identify key active motifs.

Biersack *et al.* reported that hybrid **67a**, a ferrocene-plumbagin compound, exhibited strong antiproliferative effects against 518A2 melanoma and HCT-116 colon carcinoma cells, which are responsive to plumbagin, with IC_{50} values of 0.5 μM and 4.7 μM, respectively [109]. Moreover, hybrid **67a** overcame multidrug resistance in KB-V1/Vbl cervical cancer cells, achieving an IC_{50} around 1 μM in 72 hours by using MTT assays. Hybrid **67a** induced cancer cell death (sub-G0/G1 fraction) in a time- and dose- dependent manner and significantly increased ROS levels. It also exhibited a higher affinity for linear DNA as compared to plumbagin. Although juglone showed antiproliferative activity comparable to plumbagin, its ferrocenyl derivative **67b** exhibited almost no activity. Recently, Bansode *et al.* reported that ferrocene-linked pyrimidobenzothiazoles **68a-d** and pyrimidobenzimidazole **68e** exhibited excellent potency against MDA-MB-231 and MCF-7 breast cancer cell lines, with GI_{50} values varying between 0.018 and 0.022 μM, equivalent to the standard drug doxorubicin (GI_{50}: 0.018 μM) [110].

Butler *et al.* reported that ferrocene hybrids **69a-h** exhibited moderate to strong antiproliferative effects (IC_{50}: 2.86-4.24 μM) against THP-1 leukemia cells, though they were less potent than cisplatin (IC_{50}: 1.97 μM) [111]. These hybrids inhibited tumor growth in a dose-dependent manner. SAR analysis disclosed that hybrids with a fluoro group on the phenyl ring were more effective than those with methoxy or methyl groups, likely due to fluorine's ability to form hydrogen bonds with DNA. The activity trend: F > OMe > Me suggested that electron-withdrawing groups increased the hybrids' polarity, which in turn enhanced their lipophilicity.

The ferrocene-furan hybrids **70** (IC_{50}: 0.0101-50.5 μM) demonstrated broad-spectrum anticancer activity against various cell lines, namely ovarian carcinoma (SKOV-3), cervical carcinoma (KB), non-small cell lung cancer (NCl H460), CNS cancer (SF-268), leukemia (HL60, U937, K562), melanoma (G361, SKMEL-28), colon adenocarcinoma (RKOP 27), cervical cancer (HeLa), neuroblastoma (GOTO, NB-1), MCF-7, liver cancer (HepG2), and lung fibrosarcoma (HT1080) [112, 113]. These metal complex hybrids were usually more effective than the ligands. Furthermore, their potency was comparable to that of doxorubicin (IC_{50}: 1.13-6.66 μM) in the majority of the cancer cell lines evaluated, indicating that ferrocene-furan hybrids hold potential as promising candidates for future cancer treatment development.

Recently, Mendoza-Morales *et al.* reported the synthesis and evaluation of a series of (*Z*)- and (*E*)-2-substituted-3-ferrocene-acrylonitrile hybrids **71**, which exhibited significant anticancer potential against MCF-7 and MDA-MB-231 breast cancer cell lines [114]. Among these, hybrids **71a–b** were the most active, with GI_{50} values of 11.9 and 9.1 μM, respectively, against the highly aggressive MDA-MB-231 line. The *Z*-isomers were found to be more active than their *E*-counterparts. Mechanistic investigations suggested that the anticancer activity is associated with the generation of Reactive Oxygen Species (ROS), particularly hydroxyl radicals, through Fenton-like reactions of the ferrocene moiety with endogenous hydrogen peroxide in cancer cells.

In addition to the ferrocene hybrids discussed above, numerous other reports highlight ferrocene-appended hybrids with potent anticancer activity. While a comprehensive review of all such compounds is beyond the scope of this chapter, we have focused on selected molecules whose structure-activity relationships provide valuable insights for designing more potent anticancer hybrids. These examples serve as a foundation for future development in both anticancer and antitubercular research.

CONCLUSION

Cancer remains a major cause of death worldwide, with morbidity and mortality rates expected to rise in the coming years. Despite the development of various therapeutic strategies, major challenges such as drug resistance and limited specificity of current treatments highlight the urgent need for novel chemotherapeutics that are both highly effective and selective. In this context, molecular hybridization has emerged as a powerful strategy, combining two or more biologically active scaffolds to design multitargeted compounds capable of enhanced anticancer activity.

Ferrocene derivatives have proven to be particularly promising candidates for hybrid-based drug design. Notably, the ferrocene-phenol hybrid Ferrocifen exhibits a unique mechanism of action, interacting with DNA as well as multiple proteins and enzymes, which differentiates it from conventional platinum-based anticancer agents. Over the past decade, numerous ferrocene-based hybrids, including ferrocene-amino acids, ferrocene-azoles, ferrocene-artemisinin, ferrocene-nucleic acids, and ferrocene-coumarin conjugates, have been synthesized and shown significant activity against both drug-sensitive and drug-resistant cancer cell lines. The anticancer efficacy of these hybrids is largely attributed to their ability to induce Reactive Oxygen Species (ROS), which is further enhanced when ferrocene is linked to pharmacophores such as azoles, pyrimidines, and coumarins. In addition to ROS-mediated activity, structural modifications play a crucial role in optimizing pharmacological properties. The incorporation of functional groups such as $-NH_2$, $-F$, $-OH$, and $-CF_3$ improves lipophilicity, metabolic stability, and blood-brain barrier penetration, leading to enhanced absorption, transport, and overall bioavailability. Such chemical tuning enables ferrocene hybrids to exert potent cytotoxic effects while maintaining a favorable safety profile, highlighting their potential as low-toxicity anticancer agents.

Despite these promising developments, several challenges must be addressed to enable clinical translation. Bridging the gap between encouraging *in-vitro* cytotoxicity and reproducible *in-vivo* efficacy remains a key hurdle. Furthermore, detailed studies on pharmacokinetics, pharmacodynamics, and long-term toxicity are essential to ensure therapeutic safety and efficacy. Systematic preclinical and clinical evaluations will be critical to fully realize the potential of ferrocene-based hybrids as next-generation anticancer agents. Continued research in this area is expected to provide valuable insights into the design of more effective, multitargeted, and selective anticancer therapies, ultimately advancing the fight against drug-resistant cancers.

REFERENCES

[1] Siegel RL, Miller KD, Jemal A. Cancer statistics, 2015. CA Cancer J Clin 2015; 65(1): 5-29. [http://dx.doi.org/10.3322/caac.21254] [PMID: 25559415]

[2] World Cancer Research Fund International, Available from: http://www.wcrf.org/ int/cancer-factsfigures/

[3] Romero-Canelón I, Sadler PJ. Next-generation metal anticancer complexes: multitargeting *via* redox modulation. Inorg Chem 2013; 52(21): 12276-91. [http://dx.doi.org/10.1021/ic400835n] [PMID: 23879584]

[4] Um IS, Armstrong-Gordon E, Moussa YE, Gnjidic D, Wheate NJ. Platinum drugs in the Australian cancer chemotherapy healthcare setting: Is it worthwhile for chemists to continue to develop platinums? Inorg Chim Acta 2019; 492: 177-81. [http://dx.doi.org/10.1016/j.ica.2019.04.023]

[5] Apps MG, Choi EHY, Wheate NJ. The state-of-play and future of platinum drugs. Endocr Relat Cancer 2015; 22(4): R219-33. [http://dx.doi.org/10.1530/ERC-15-0237] [PMID: 26113607]

[6] Galanski MS, Jakupec MA, Keppler BK. Update of the preclinical situation of anticancer platinum complexes: novel design strategies and innovative analytical approaches. Curr Med Chem 2005; 12(18): 2075-94. [http://dx.doi.org/10.2174/0929867054637626] [PMID: 16101495]

[7] Ali I, Wani WA, Saleem K, Haque A. Platinum compounds: a hope for future cancer chemotherapy. Anticancer Agents Med Chem 2013; 13(2): 296-306. [http://dx.doi.org/10.2174/1871520611313020016] [PMID: 22583420]

[8] Ott I, Gust R. Non platinum metal complexes as anti-cancer drugs. Arch Pharm (Weinheim) 2007; 340(3): 117-26. [http://dx.doi.org/10.1002/ardp.200600151] [PMID: 17315259]

[9] Kicic A, Chua ACG, Baker E. The desferrithiocin (DFT) class of iron chelators: potential as

antineoplastic agents Anticancer Drug Des 2001; 16(4-5): 195-207.
[PMID: 12049478]

[10] Weinberg ED. Development of clinical methods of iron deprivation for suppression of neoplastic and infectious diseases. Cancer Invest 1999; 17(7): 507-13.
[http://dx.doi.org/10.3109/07357909909032860] [PMID: 10518195]

[11] Mir LM, Tounekti O, Orlowski S. Bleomycin: Revival of an old drug. Gen Pharmacol 1996; 27(5): 745-8.
[http://dx.doi.org/10.1016/0306-3623(95)02101-9] [PMID: 8842674]

[12] Umezawa H. Bleomycin and other antitumor antibiotics of high molecular weight. Antimicrob Agents Chemother 1965; 5: 1079-85.
[PMID: 5883403]

[13] Yu Y, Gutierrez E, Kovacevic Z, *et al.* Iron chelators for the treatment of cancer. Curr Med Chem 2012; 19(17): 2689-702.
[http://dx.doi.org/10.2174/092986712800609706] [PMID: 22455580]

[14] Rauf U, Shabir G, Bukhari S, Albericio F, Saeed A. Contemporary developments in ferrocene chemistry: physical, chemical, biological and industrial aspects. Molecules 2023; 28(15): 5765.
[http://dx.doi.org/10.3390/molecules28155765] [PMID: 37570735]

[15] Patra M, Gasser G. The medicinal chemistry of ferrocene and its derivatives. Nat. Rev. Chem. 2017, 1, 0066.
[http://dx.doi.org/10.1038/s41570-017-0066]

[16] Jaouen G, Vessières A, Top S. Ferrocifen type anti cancer drugs. Chem Soc Rev 2015; 44(24): 8802-17.
[http://dx.doi.org/10.1039/C5CS00486A] [PMID: 26486993]

[17] Ornelas C, Astruc D. Ferrocene-based drugs, delivery nanomaterials and fenton mechanism: State of the art, recent developments and prospects. Pharmaceutics 2023; 15(8): 2044.
[http://dx.doi.org/10.3390/pharmaceutics15082044] [PMID: 37631259]

[18] (a) Gu H, Mu S, Qiu G, *et al.* Redox-stimuli-responsive drug delivery systems with supramolecular ferrocenyl-containing polymers for controlled release. Coord Chem Rev 2018; 364: 51-85.
[http://dx.doi.org/10.1016/j.ccr.2018.03.013]
(b) Reshetnikov V, Özkan HG, Daum S, *et al.* *N*-Alkylaminoferrocene-based prodrugs targeting mitochondria of cancer cells. Molecules 2020; 25(11): 2545.
[http://dx.doi.org/10.3390/molecules25112545] [PMID: 32486084]

[19] Sharma A, Rana R, Kumar N, *et al.* Remarkable contribution of ferrocene as anticancer agent: An updated review (2020–2024). Inorg Chem Commun 2025; 175: 114155.
[http://dx.doi.org/10.1016/j.inoche.2025.114155]

[20] Juan CA, Pérez de la Lastra JM, Plou FJ, Pérez-Lebeña E. The chemistry of Reactive Oxygen Species (ROS) revisited: Outlining their role in biological macromolecules (DNA, lipids and proteins) and induced pathologies. Int J Mol Sci 2021; 22(9): 4642.
[http://dx.doi.org/10.3390/ijms22094642] [PMID: 33924958]

[21] Jomova K, Raptova R, Alomar SY, *et al.* Reactive oxygen species, toxicity, oxidative stress, and antioxidants: chronic diseases and aging. Arch Toxicol 2023; 97(10): 2499-574.
[http://dx.doi.org/10.1007/s00204-023-03562-9] [PMID: 37597078]

[22] Arfin S, Jha NK, Jha SK, *et al.* Oxidative stress in cancer cell metabolism. Antioxidants 2021; 10(5): 642.
[http://dx.doi.org/10.3390/antiox10050642] [PMID: 33922139]

[23] Sharma A, Rana R, Kumar N, *et al.* Ferrocene-based compounds: promising anticancer and antimalarial agents in modern therapeutics. Med Chem Res 2025; 34(6): 1177-99.
[http://dx.doi.org/10.1007/s00044-025-03408-2]

[24] Rescifina A, Zagni C, Varrica MG, Pistarà V, Corsaro A. Recent advances in small organic molecules as DNA intercalating agents: Synthesis, activity, and modeling. Eur J Med Chem 2014; 74: 95-115. [http://dx.doi.org/10.1016/j.ejmech.2013.11.029] [PMID: 24448420]

[25] Zhang Y, Li S, Zhou H, Wang Y, Wang L. Ferrocene-quinoline hybrids: Synthesis, DNA binding, and cytotoxic activity. Dalton Trans 2020; 49: 12292-305.

[26] Fouda AM, Abdelrahman MH, Ali I. Redox-active ferrocenyl conjugates: Role of Fe(II)/Fe(III) in biological reactivity. Coord Chem Rev 2021; 447: 214143.

[27] Hossion AM, Tanaka R, Wakimoto M, Kouno I, Tanaka T. DNA topoisomerase II inhibition by ferrocene-quinoline derivatives. J Med Chem 2016; 59: 1055-65.

[28] Huo J, Liu J, Wu W, *et al.* Ferrocenyl derivatives as DNA-targeted topoisomerase II inhibitors. Bioorg Med Chem 2018; 26: 2891-9.

[29] Trachootham D, Alexandre J, Huang P. Targeting cancer cells by ROS-mediated mechanisms: a radical therapeutic approach? Nat Rev Drug Discov 2009; 8(7): 579-91. [http://dx.doi.org/10.1038/nrd2803] [PMID: 19478820]

[30] Yang H, Villani RM, Wang H, *et al.* The role of cellular redox imbalance in cancer progression and therapy. Redox Biol 2018; 14: 457-73.

[31] Jordan MA, Wilson L. Microtubules as a target for anticancer drugs. Nat Rev Cancer 2004; 4(4): 253-65. [http://dx.doi.org/10.1038/nrc1317] [PMID: 15057285]

[32] de Oliveira KM, Guedes da Silva CC, Pombeiro AJL. Ferrocenyl chalcone derivatives: Tubulin inhibition and antiproliferative activity. J Inorg Biochem 2020; 210: 111148.

[33] Huo J, Liu J, Wu W, *et al.* Ferrocenyl quinoline hybrids as tubulin-targeting anticancer agents. Eur J Med Chem 2018; 150: 543-54.

[34] Rescifina A, Zagni C, Vessières A, Jaouen G. Ferrocene hybrids in cancer therapy: ROS, DNA damage, and cytoskeletal disruption. Future Med Chem 2014; 6: 1113-32.

[35] Zhang C, Sun Y, Huang J, *et al.* Multifunctional ferrocenyl hybrids: From ROS generation to DNA and tubulin targeting. Eur J Med Chem 2020; 188: 111953.

[36] Kim CH, Axup JY, Schultz PG. Protein conjugation with genetically encoded unnatural amino acids. Curr Opin Chem Biol 2013; 17(3): 412-9. [http://dx.doi.org/10.1016/j.cbpa.2013.04.017] [PMID: 23664497]

[37] Blaskovich MAT. Unusual amino acids in medicinal chemistry. J Med Chem 2016; 59(24): 10807-36. [http://dx.doi.org/10.1021/acs.jmedchem.6b00319] [PMID: 27589349]

[38] Goswami TK, Gadadhar S, Gole B, Karande AA, Chakravarty AR. Photocytotoxicity of copper(II) complexes of curcumin and N-ferrocenylmethyl-l-amino acids. Eur J Med Chem 2013; 63: 800-10. [http://dx.doi.org/10.1016/j.ejmech.2013.03.026] [PMID: 23584543]

[39] Goswami TK, Gadadhar S, Balaji B, Gole B, Karande AA, Chakravarty AR. Ferrocenyl- L -amino acid copper(II) complexes showing remarkable photo-induced anticancer activity in visible light. Dalton Trans 2014; 43(31): 11988-99. [http://dx.doi.org/10.1039/C4DT01348D] [PMID: 24971754]

[40] Goswami TK, Gadadhar S, Roy M, Nethaji M, Karande AA, Chakravarty AR. Ferrocene-conjugated copper(II) complexes of L -methionine and phenanthroline bases: Synthesis, structure, and photocytotoxic activity. Organometallics 2012; 31(8): 3010-21. [http://dx.doi.org/10.1021/om201102k]

[41] Butler WE, Kelly PN, Harry AG, *et al.* The synthesis, structural characterization and biological evaluation of *N*-(ferrocenylmethyl amino acid) fluorinated benzene-carboxamide derivatives as potential anticancer agents. Appl Organomet Chem 2013; 27(6): 361-5.

[http://dx.doi.org/10.1002/aoc.2994]

[42] Kovačević M, Kodrin I, Cetina M, *et al.* The conjugates of ferrocene-1,1′-diamine and amino acids. A novel synthetic approach and conformational analysis. Dalton Trans 2015; 44(37): 16405-20. [http://dx.doi.org/10.1039/C5DT01610J] [PMID: 26308626]

[43] Tauchman J, Süss-Fink G, Štěpnička P, Zava O, Dyson PJ. Arene ruthenium complexes with phosphinoferrocene amino acid conjugates: Synthesis, characterization and cytotoxicity. J Organomet Chem 2013; 723: 233-8. [http://dx.doi.org/10.1016/j.jorganchem.2012.10.009]

[44] Harry AG, Butler WE, Manton JC, *et al.* The synthesis, structural characterization and *in vitro* anticancer activity of novel 1-alkyl-1′-N-meta-(ferrocenyl) benzoyl dipeptide esters and novel 1-alkyl-1′-N-ortho-(ferrocenyl) benzoyl dipeptide esters. J Organomet Chem 2014; 766: 1-12. [http://dx.doi.org/10.1016/j.jorganchem.2014.04.027]

[45] Harry AG, Butler WE, Manton JC, *et al.* The synthesis, structural characterization and *in vitro* anti-cancer activity of novel 1-alkyl-1′-N-para-(ferrocenyl) benzoyl dipeptide esters. J Organomet Chem 2014; 757: 28-35. [http://dx.doi.org/10.1016/j.jorganchem.2014.01.031]

[46] Zhang L, Xu Z. Coumarin-containing hybrids and their anticancer activities. Eur J Med Chem 2019; 181: 111587. [http://dx.doi.org/10.1016/j.ejmech.2019.111587] [PMID: 31404864]

[47] Tasior M, Kim D, Singha S, Krzeszewski M, Ahn KH, Gryko DT. π-Expanded coumarins: synthesis, optical properties and applications. J Mater Chem C Mater Opt Electron Devices 2015; 3(7): 1421-46. [http://dx.doi.org/10.1039/C4TC02665A]

[48] Mbaba M, Mabhula AN, Boel N, *et al.* Ferrocenyl and organic novobiocin derivatives: Synthesis and their *in vitro* biological activity. J Inorg Biochem 2017; 172: 88-93. [http://dx.doi.org/10.1016/j.jinorgbio.2017.04.014] [PMID: 28441548]

[49] Mbaba M, de la Mare JA, Sterrenberg JN, *et al.* Novobiocin–ferrocene conjugates possessing anticancer and antiplasmodial activity independent of HSP90 inhibition. J Biol Inorg Chem 2019; 24(2): 139-49. [http://dx.doi.org/10.1007/s00775-018-1634-9] [PMID: 30542925]

[50] Wei JN, Jia ZD, Zhou YQ, *et al.* Synthesis, characterization and antitumor activity of novel ferrocene-coumarin conjugates. J Organomet Chem 2019; 902: 120968. [http://dx.doi.org/10.1016/j.jorganchem.2019.120968]

[51] Pérès B, Nasr R, Zarioh M, *et al.* Ferrocene-embedded flavonoids targeting the Achilles heel of multidrug-resistant cancer cells through collateral sensitivity. Eur J Med Chem 2017; 130: 346-53. [http://dx.doi.org/10.1016/j.ejmech.2017.02.064] [PMID: 28273561]

[52] Monserrat JP, Tiwari KN, Quentin L, *et al.* Ferrocenyl flavonoid-induced morphological modifications of endothelial cells and cytotoxicity against B16 murine melanoma cells. J Organomet Chem 2013; 734: 78-85. [http://dx.doi.org/10.1016/j.jorganchem.2012.12.031]

[53] Kowalski K, Szczupak Ł, Oehninger L, *et al.* Ferrocenyl derivatives of pterocarpene and coumestan: Synthesis, structure and anticancer activity studies. J Organomet Chem 2014; 772-773: 49-59. [http://dx.doi.org/10.1016/j.jorganchem.2014.08.027]

[54] Taner B, Mahmood Mahmood AH, Karakurt S, Altuner EE, Dar UA. Synthesis, characterization and biological evaluation of new Coumarin-Ferrocene conjugates as antitumor agents. Inorg Chim Acta 2025; 578: 122534. [http://dx.doi.org/10.1016/j.ica.2024.122534]

[55] Dexamethasone alone or in combination with ondansetron for the prevention of delayed nausea and vomiting induced by chemotherapy. N Engl J Med 2000; 342(21): 1554-9.

[http://dx.doi.org/10.1056/NEJM200005253422102] [PMID: 10824073]

[56] Narváez-Pita X, Rheingold AL, Meléndez E. Ferrocene-steroid conjugates: Synthesis, structure and biological activity. J Organomet Chem 2017; 846: 113-20. [http://dx.doi.org/10.1016/j.jorganchem.2017.06.004]

[57] Manosroi J, Rueanto K, Boonpisuttinant K, *et al.* Novel ferrocenic steroidal drug derivatives and their bioactivities. J Med Chem 2010; 53(10): 3937-43. [http://dx.doi.org/10.1021/jm901866m] [PMID: 20408531]

[58] Vera J, Gao LM, Santana A, Matta J, Meléndez E. Vectorized ferrocenes with estrogens and vitamin D2: synthesis, cytotoxic activity and docking studies. Dalton Trans 2011; 40(37): 9557-65. [http://dx.doi.org/10.1039/c1dt10995b] [PMID: 21850331]

[59] Kuznetsov AE. Azoles: synthesis, properties, applications and perspectives. BoD – Books on Demand 2021. [http://dx.doi.org/10.5772/intechopen.87765]

[60] Prasher P, Sharma M, Zacconi F, *et al.* Synthesis and anticancer properties of 'azole'based chemotherapeutics as emerging chemical moieties: A comprehensive review. Curr Org Chem 2021; 25(6): 654-68. [http://dx.doi.org/10.2174/1385272824999200820152501]

[61] Snegur LV, Nekrasov YS, Sergeeva NS, *et al.* Ferrocenylalkyl azoles: bioactivity, synthesis, structure. Appl Organomet Chem 2008; 22(2): 139-47. [http://dx.doi.org/10.1002/aoc.1362]

[62] Alghamdi SS, Suliman RS, Almutairi K, Kahtani K, Aljatli D. Imidazole as a promising medicinal scaffold: current status and future direction. Drug Des Devel Ther 2021; 15: 3289-312. [http://dx.doi.org/10.2147/DDDT.S307113] [PMID: 34354342]

[63] Muenzner JK, Biersack B, Albrecht A, *et al.* Ferrocenyl-coupled N-heterocyclic carbene complexes of gold(I): A successful approach to multinuclear anticancer drugs. Chemistry 2016; 22(52): 18953-62. [http://dx.doi.org/10.1002/chem.201604246] [PMID: 27761940]

[64] Tabrizi L, Chiniforoshan H. The cytotoxicity and mechanism of action of new multinuclear Scaffold Au^{III}, Pd^{II} pincer complexes containing a bis(diphenylphosphino) ferrocene/non-ferrocene ligand. Dalton Trans 2017; 46(41): 14164-73. [http://dx.doi.org/10.1039/C7DT02887C] [PMID: 28984321]

[65] Kanoh K, Kohno S, Katada J, Takahashi J, Uno I. (-)-Phenylahistin arrests cells in mitosis by inhibiting tubulin polymerization. J Antibiot (Tokyo) 1999; 52(2): 134-41. [http://dx.doi.org/10.7164/antibiotics.52.134] [PMID: 10344567]

[66] Wieczorek A, Błauż A, Zakrzewski J, Rychlik B, Plażuk D. Ferrocenyl 2,5-Piperazinediones as Tubulin-Binding Organometallic $ABCB_1$ and $ABCG_2$ Inhibitors Active against MDR Cells. ACS Med Chem Lett 2016; 7(6): 612-7. [http://dx.doi.org/10.1021/acsmedchemlett.6b00046] [PMID: 27326336]

[67] Atmaca H, Özkan AN, Zora M. Novel ferrocenyl pyrazoles inhibit breast cancer cell viability *via* induction of apoptosis and inhibition of PI3K/Akt and ERK1/2 signaling. Chem Biol Interact 2017; 263: 28-35. [http://dx.doi.org/10.1016/j.cbi.2016.12.010] [PMID: 27989600]

[68] Gupta A, Das R, Chamoli A, *et al.* A Series of ferrocene-containing pyrazolo[1,5-*a*]pyrimidines induce a strong antiproliferative effect against oral cancer cells. Organometallics 2022; 41(16): 2365-78. [http://dx.doi.org/10.1021/acs.organomet.2c00348]

[69] Li B, Zhu X, Guo Y, Ren YP, Jia ZD, Wei JN, Fu DX, Xu Y, Hao XQ. Synthesis and properties of m-ferrocenylbenzoylthiadiazole derivatives. Appl. Organomet. Chem. 2018, 32, e4265 (1-8). [http://dx.doi.org/10.1002/aoc.4265]

[70] Dheer D, Singh V, Shankar R. Medicinal attributes of 1,2,3-triazoles: Current developments. Bioorg Chem 2017; 71: 30-54.
[http://dx.doi.org/10.1016/j.bioorg.2017.01.010] [PMID: 28126288]

[71] Sharma A, Agrahari AK, Rajkhowa S, Tiwari VK. Emerging impact of triazoles as anti-tubercular agent. Eur J Med Chem 2022; 238: 114454.
[http://dx.doi.org/10.1016/j.ejmech.2022.114454] [PMID: 35597009]

[72] Plażuk D, Rychlik B, Błauż A, Domagała S. Synthesis, electrochemistry and anticancer activity of novel ferrocenyl phenols prepared *via* azide-alkyne 1,3-cycloaddition reaction. J Organomet Chem 2012; 715: 102-12.
[http://dx.doi.org/10.1016/j.jorganchem.2012.05.042]

[73] Ghosh N, Chaki R, Mandal V, Mandal SC. COX-2 as a target for cancer chemotherapy. Pharmacol Rep 2010; 62(2): 233-44.
[http://dx.doi.org/10.1016/S1734-1140(10)70262-0] [PMID: 20508278]

[74] Ren SZ, Wang ZC, Zhu D, *et al.* Design, synthesis and biological evaluation of novel ferrocene-pyrazole derivatives containing nitric oxide donors as COX-2 inhibitors for cancer therapy. Eur J Med Chem 2018; 157: 909-24.
[http://dx.doi.org/10.1016/j.ejmech.2018.08.048] [PMID: 30149323]

[75] Shivatare SS, Shivatare VS, Wong CH. Glycoconjugates: synthesis, functional studies, and therapeutic developments. Chem Rev 2022; 122(20): 15603-71.
[http://dx.doi.org/10.1021/acs.chemrev.1c01032] [PMID: 36174107]

[76] Casas-Solvas JM, Ortiz-Salmerón E, Giménez-Martínez JJ, *et al.* Ferrocene-carbohydrate conjugates as electrochemical probes for molecular recognition studies. Chemistry 2009; 15(3): 710-25.
[http://dx.doi.org/10.1002/chem.200800927] [PMID: 19053085]

[77] Trivedi R, Deepthi SB, Giribabu L, *et al.* Synthesis, crystal structure, electronic spectroscopy, electrochemistry and biological studies of carbohydrate containing ferrocene amides. Appl Organomet Chem 2012; 26(7): 369-76.
[http://dx.doi.org/10.1002/aoc.2874]

[78] Trivedi R, Deepthi SB, Giribabu L, *et al.* Synthesis, crystal structure, electronic spectroscopy, electrochemistry and biological studies of ferrocene–carbohydrate conjugates. Eur J Inorg Chem 2012; 2012(13): 2267-77.
[http://dx.doi.org/10.1002/ejic.201200038]

[79] Deepthi SB, Trivedi R, Giribabu L, Sujitha P, Kumar CG. Effect of amide-triazole linkers on the electrochemical and biological properties of ferrocene-carbohydrate conjugates. Dalton Trans 2013; 42(4): 1180-90.
[http://dx.doi.org/10.1039/C2DT31927F] [PMID: 23111690]

[80] Daum S, Toms J, Reshetnikov V, *et al.* Identification of boronic acid derivatives as an active form of *N* -alkylaminoferrocene-based anticancer prodrugs and their radiolabeling with ^{18} F. Bioconjug Chem 2019; 30(4): 1077-86.
[http://dx.doi.org/10.1021/acs.bioconjchem.9b00019] [PMID: 30768258]

[81] Sharma V, Chitranshi N, Agarwal AK. Significance and biological importance of pyrimidine in the microbial world. Int J Med Chem 2014; 2014: 1-31.
[http://dx.doi.org/10.1155/2014/202784] [PMID: 25383216]

[82] Skiba J, Kowalski K, Prochnicka A, *et al.* Metallocene-uracil conjugates: Synthesis and biological evaluation of novel mono-, di- and tri-nuclear systems. J Organomet Chem 2015; 782: 52-61.
[http://dx.doi.org/10.1016/j.jorganchem.2014.11.017]

[83] Skiba J, Karpowicz R, Szabó I, Therrien B, Kowalski K. Synthesis and anticancer activity studies of ferrocenyl-thymine-3,6-dihydro-2H-thiopyranes – A new class of metallocene-nucleobase derivatives. J Organomet Chem 2015; 794: 216-22.

[http://dx.doi.org/10.1016/j.jorganchem.2015.07.012]

[84] Sansook S, Lineham E, Hassell-Hart S, *et al.* Probing the anticancer action of novel ferrocene analogues of MNK Inhibitors. Molecules 2018; 23(9): 2126.
[http://dx.doi.org/10.3390/molecules23092126] [PMID: 30142961]

[85] Simenel AA, Morozova EA, Snegur LV, *et al.* Simple route to ferrocenylalkyl nucleobases. Antitumor activity *in vivo*. Appl Organomet Chem 2009; 23(6): 219-24.
[http://dx.doi.org/10.1002/aoc.1500]

[86] Guo Y, Wang SQ, Ding ZQ, Zhou J, Ruan BF. Synthesis, characterization and antitumor activity of novel ferrocene bisamide derivatives containing pyrimidine-moiety. J Organomet Chem 2017; 851: 150-9.
[http://dx.doi.org/10.1016/j.jorganchem.2017.09.032]

[87] Marella A, Tanwar OP, Saha R, *et al.* Quinoline: A versatile heterocyclic. Saudi Pharm J 2013; 21(1): 1-12.
[http://dx.doi.org/10.1016/j.jsps.2012.03.002] [PMID: 23960814]

[88] Ilakiyalakshmi M, Arumugam Napoleon A. Review on recent development of quinoline for anticancer activities. Arab J Chem 2022; 15(11): 104168.
[http://dx.doi.org/10.1016/j.arabjc.2022.104168]

[89] Jain S, Chandra V, Kumar Jain P, Pathak K, Pathak D, Vaidya A. Comprehensive review on current developments of quinoline-based anticancer agents. Arab J Chem 2019; 12(8): 4920-46.
[http://dx.doi.org/10.1016/j.arabjc.2016.10.009]

[90] Stringer T, De Kock C, Guzgay H, *et al.* Mono- and multimeric ferrocene congeners of quinoline-based polyamines as potential antiparasitics. Dalton Trans 2016; 45(34): 13415-26.
[http://dx.doi.org/10.1039/C6DT02685K] [PMID: 27485032]

[91] N'Da DD, Smith PJ. Synthesis, *in vitro* antiplasmodial and antiproliferative activities of a series of quinoline–ferrocene hybrids. Med Chem Res 2014; 23(3): 1214-24.
[http://dx.doi.org/10.1007/s00044-013-0748-4]

[92] Maračić S, Lapić J, Djaković S, *et al.* Quinoline and ferrocene conjugates: Synthesis, computational study and biological evaluations. Appl Organomet Chem 2019; 33(1): e4628.
[http://dx.doi.org/10.1002/aoc.4628]

[93] Kocsis L, Szabó I, Bősze S, Jernei T, Hudecz F, Csámpai A. Synthesis, structure and *in vitro* cytostatic activity of ferrocene—Cinchona hybrids. Bioorg Med Chem Lett 2016; 26(3): 946-9.
[http://dx.doi.org/10.1016/j.bmcl.2015.12.059] [PMID: 26739780]

[94] Podolski-Renić A, Bősze S, Dinić J, Kocsis L, Hudecz F, Csámpai A, Pešić M. Ferrocene-cinchona hybrids with triazolyl-chalcone linkers act as pro-oxidants and sensitize human cancer cell lines to paclitaxel. Metallomics : integrated biometal science, 2017, 9, 1132–1141.

[95] Krishna S, Bustamante L, Haynes RK, Staines HM. Artemisinins: their growing importance in medicine. Trends Pharmacol Sci 2008; 29(10): 520-7.
[http://dx.doi.org/10.1016/j.tips.2008.07.004] [PMID: 18752857]

[96] Posadino AM, Giordo R, Pintus G, *et al.* Medicinal and mechanistic overview of artemisinin in the treatment of human diseases. Biomed Pharmacother 2023; 163: 114866.
[http://dx.doi.org/10.1016/j.biopha.2023.114866] [PMID: 37182516]

[97] Gao F, Sun Z, Kong F, Xiao J. Artemisinin-derived hybrids and their anticancer activity. Eur J Med Chem 2020; 188: 112044.
[http://dx.doi.org/10.1016/j.ejmech.2020.112044] [PMID: 31945642]

[98] Reiter C, Fröhlich T, Zeino M, *et al.* New efficient artemisinin derived agents against human leukemia cells, human cytomegalovirus and Plasmodium falciparum: 2nd generation 1,2,4-trioxane-ferrocene hybrids. Eur J Med Chem 2015; 97: 164-72.
[http://dx.doi.org/10.1016/j.ejmech.2015.04.053] [PMID: 25965779]

[99] Reiter C, Çapcı Karagöz A, Fröhlich T, *et al.* Synthesis and study of cytotoxic activity of 1,2,4-trioxane- and egonol-derived hybrid molecules against Plasmodium falciparum and multidrug-resistant human leukemia cells. Eur J Med Chem 2014; 75: 403-12.
[http://dx.doi.org/10.1016/j.ejmech.2014.01.043] [PMID: 24561670]

[100] Top S, Tang J, Vessières A, Carrez D, Provot C, Jaouen G. Ferrocenyl hydroxytamoxifen: a prototype for a new range of oestradiol receptor site-directed cytotoxics. Chem Commun (Camb) 1996; (8): 955-6.
[http://dx.doi.org/10.1039/CC9960000955]

[101] Top S, Dauer B, Vaissermann J, Jaouen G. Facile route to ferrocifen, 1-[4-(2-dimethylaminoethoxy)]-1-(phenyl-2-ferrocenyl-but-1-ene), first organometallic analogue of tamoxifen, by the McMurry reaction. J Organomet Chem 1997; 541(1-2): 355-61.
[http://dx.doi.org/10.1016/S0022-328X(97)00086-7]

[102] Vessières A. Metal carbonyl tracers and the ferrocifen family: Two facets of bioorganometallic chemistry. J Organomet Chem 2013; 734: 3-16.
[http://dx.doi.org/10.1016/j.jorganchem.2012.12.020]

[103] Nguyen A, Vessières A, Hillard EA, Top S, Pigeon P, Jaouen G. Ferrocifens and ferrocifenols as new potential weapons against breast cancer. Chimia (Aarau) 2007; 61(11): 716-24.
[http://dx.doi.org/10.2533/chimia.2007.716]

[104] Hillard E, Vessières A, Le Bideau F, *et al.* A series of unconjugated ferrocenyl phenols: prospects as anticancer agents. ChemMedChem 2006; 1(5): 551-9.
[http://dx.doi.org/10.1002/cmdc.200500035] [PMID: 16892391]

[105] González-Pelayo S, López E, Borge J, de-los-Santos-Álvarez N, López LA. Ferrocene-decorated phenol derivatives by trapping *ortho* -quinone methide intermediates with ferrocene. Eur J Org Chem 2018; 2018(22): 2858-62.
[http://dx.doi.org/10.1002/ejoc.201800396]

[106] González-Pelayo S, López E, Borge J, De-los-Santos-Álvarez N, López LA. Trapping *para*-quinone methide intermediates with ferrocene: Synthesis and preliminary biological evaluation of new phenol-ferrocene conjugates. Molecules 2018; 23(6): 1335.
[http://dx.doi.org/10.3390/molecules23061335] [PMID: 29865205]

[107] Plażuk D, Wieczorek A, Ciszewski WM, *et al.* Synthesis and *in vitro* biological evaluation of ferrocenyl side-chain-functionalized paclitaxel derivatives. ChemMedChem 2017; 12(22): 1882-92.
[http://dx.doi.org/10.1002/cmdc.201700576] [PMID: 28941201]

[108] Wang Y, Pigeon P, Top S, McGlinchey MJ, Jaouen G. Organometallic antitumor compounds: ferrocifens as precursors to quinone methides. Angew Chem Int Ed 2015; 54(35): 10230-3.
[http://dx.doi.org/10.1002/anie.201503048] [PMID: 26179051]

[109] Spoerlein-Guettler C, Mahal K, Schobert R, Biersack B. Ferrocene and (arene)ruthenium(II) complexes of the natural anticancer naphthoquinone plumbagin with enhanced efficacy against resistant cancer cells and a genuine mode of action. J Inorg Biochem 2014; 138: 64-72.
[http://dx.doi.org/10.1016/j.jinorgbio.2014.04.020] [PMID: 24907976]

[110] Bansode P, Pore D, Tayade S, Patil S, Choudhari P, Rashinkar G. Remarkable anti-breast cancer activity and molecular docking studies of ferrocene tethered pyrimidobenzothiazoles and pyrimidobenzimidazoles. Results Chem 2023; 5: 100758.
[http://dx.doi.org/10.1016/j.rechem.2022.100758]

[111] Asghar F, Badshah A, Lal B, Zubair S, Fatima S, Butler IS. Facile synthesis of fluoro, methoxy, and methyl substituted ferrocene-based urea complexes as potential therapeutic agents. Bioorg Chem 2017; 72: 215-27.
[http://dx.doi.org/10.1016/j.bioorg.2017.04.016] [PMID: 28482262]

[112] Abd-Elzaher MM, Labib AA, Mousa HA, Moustafa SA, Abdallah MM. Synthesis, characterization

and cytotoxic activity of ferrocenyl hydrazone complexes containing a furan moiety. Res Chem Intermed 2014; 40(5): 1923-36.
[http://dx.doi.org/10.1007/s11164-013-1090-7]

[113] Tombul M, Bulut A, Turk M, Ucar B, Isilar O. Synthesis and biological activity of ferrocenyl furoyl derivatives. *Inorg. Nano-Metal.* Chem 2017; 47: 865-9.

[114] Mendoza-Morales WO, Rodríguez E, González A, *et al.* Synthesis, structure, electrochemistry, and *in vitro* anticancer and anti-migratory activities of (*Z*)- and (*E*)-2-substituted-3-ferrocene-acrylonitrile hybrids and their derivatives. Molecules 2025; 30(13): 2835.
[http://dx.doi.org/10.3390/molecules30132835] [PMID: 40649349]

CHAPTER 7

Fascinations in Biocoordination Chemistry involved in Metallodrugs and Herbomineral Formulations: Harnessing Indian and Global Biomedical Knowledge System

A.P. Mishra[1,*] and **Vinayshri Tripathi**[1]

[1] *Bio-inorganic Research Lab, Department of Chemistry, Dr. Hari Singh Gour Vishwavidyalaya (A Central University), Sagar, India*

Abstract: This chapter is an attempt to explore and integrate the world's oldest biomedical system and experiential knowledge *i.e.,* the Ayurvedic metallodrugs and herbomineral formulations, with modern evidence-based pharmaceutical/therapeutic sciences. It focuses on the application of coordination chemistry, chelation, and metal-mediated biosyntheses, bio-transport, and drug delivery systems. Key themes in this chapter include the relevance of Ayurvedic metallurgical techniques, biocompatible processing in coordination chemistry, design of bioinspired metal-ligand complexes, microbe-catalysed herbomineral preparations, and the development of metallodrugs with improved biodistribution/assimilation profiles. This substantiates the modules and models, now practiced as ADMET pathways, to address issues involved in Pharmacokinetics and Pharmacodynamics. The chapter discusses how Ayurvedic chelation, using herbal APIs as ligands, aligns with contemporary chelation approaches and highlights modern analytical methods; this validates the standardization and safety of these metallodrugs. Furthermore, it addresses the challenges in integrating various world therapies and practices, including Ayurvedic, with current regulatory standards, emphasizing the need for collaborations among chemists, pharmacologists, and traditional practitioners. Examining these intersections, this chapter establishes a foundation for developing next-generation metallodrugs inspired by Ayurvedic principles. The convergence of Ayurveda with modern pharmaceutical practices offers a holistic framework, providing safe, accessible, and effective therapeutic options. Issues relating to retro-pharmacology and re-purposed drugs shall also be discussed in principle. This interdisciplinary approach not only enhances drug efficacy and safety but also enriches global healthcare by bridging ancient wisdom with contemporary science, demonstrating the potential for Ayurveda-inspired metallodrugs to address complex health challenges in a scientifically robust and culturally sensitive manner.

[*] **Corresponding author A.P. Mishra:** Bio-inorganic Research Lab, Department of Chemistry, Dr. Hari Singh Gour Vishwavidyalaya (A Central University), Sagar, India; E-mail: apmishrasagar@gmail.com

Abhay Nanda Srivastva (Ed.)
All rights reserved-© 2026 Bentham Science Publishers

Keywords: Ayurvedic metallodrugs, Bhasma, Bioavailability, Chelation therapy, Metal-ligand complexes, Nano-scaled metal drugs nanoparticles, Traditional medicine integration.

INTRODUCTION

The integration of Indian Knowledge Systems (IKS), particularly Ayurveda, with contemporary scientific methods opens a promising avenue for holistic healthcare innovation. Ayurveda, a system of medicine practiced in India for over 5,000 years, emphasizes achieving equilibrium within the body and between individuals and their environment. A unique aspect of Ayurveda is its therapeutic applications; one of them is by using metals, *viz.* gold, silver, copper, zinc, and iron as bioactive agents and herbomineral chelates. Through highly specific metallurgical processes, including Bhasmikaran (calcination) and Shodhana (purification), these raw metals are transformed into finely processed and stable medicinal formulations known as bhasmas (assaimalable metal oxides), tested qualitatively as nano-scale particles, improved bioavailability, and controlled reactivity. These metallodrugs are essential in Ayurveda for addressing a range of ailments, from inflammation and infections to immune modulation and tissue regeneration [1]. Recent research highlights the value of IKS, suggesting that Ayurvedic metallurgical principles align well with modern pharmacological requirements for stability, selectivity, and bioavailability, supporting the continued relevance of these practices.

Biocoordination Chemistry: A Conceptual Framework

Biocoordination chemistry may be broadly defined as the study of coordination interactions between biologically relevant metal ions and organic or biomolecular ligands, leading to complexes that exhibit distinct physicochemical and biological properties. Within the Ayurvedic tradition, this concept is implicitly applied in bhasma and herbomineral formulations, where metals interact with phytoconstituents such as polyphenols, flavonoids, and organic acids to generate stable complexes (*e.g.*, copper–tannin, zinc–polyphenol, or iron–ascorbate chelates). These complexes are believed to enhance therapeutic efficacy, reduce toxicity, and improve assimilation into biological systems [1].

In contemporary medicine, the same principle governs the design of metallodrugs, where metal ions coordinate with biomolecular targets such as nucleic acids and proteins. A classic example is cisplatin, where platinum coordinates to DNA bases to inhibit replication, while more recent research explores ruthenium complexes interacting with transferrin or thiol-containing proteins [6]. These interactions exemplify how metal–ligand chemistry can be tuned to achieve selectivity, controlled reactivity, and therapeutic precision.

Thus, the Ayurvedic paradigm of metal–herb synergy and the pharmaceutical paradigm of metal–biomolecule interaction converge under the umbrella of biocoordination chemistry. This unified framework provides the necessary theoretical grounding for subsequent discussions, highlighting how ancient practices can inform modern coordination-based drug discovery and delivery.To bridge the gap between these two systems and provide a clear framework for this chapter, the following Table **1** offers a comparative analysis of the fundamental principles underlying both Ayurvedic and modern metallodrugs. This comparison highlights their shared reliance on biocoordination chemistry to achieve therapeutic outcomes.

Table 1. Comparative analysis of ayurvedic and modern metallodrugs.

System	Example(s)	Coordination Chemistry Aspect	Pharmacological Implication
Ayurvedic Metallodrugs (Bhasmas)	Yashad Bhasma (ZnO nanoparticles), Tamra Bhasma (Cu-based), Swarna Bhasma (Au-based)	Stabilized by **polyphenols, flavonoids, fulvic acid** (O-donor ligands); surface modification *via* repeated *Shodhana/Bhavana*; metal-oxide core with organic coating	Detoxification, immunomodulation, antioxidant, antidiabetic, wound healing
Modern Metallodrugs	Cisplatin ($Pt(NH_3)_2Cl_2$), Ruthenium(III) complexes, Platinum-based drugs, MOFs	Coordination with **N-donor ligands** (amines, DNA bases), **S-donor** (proteins, thiols), **O-donor** (carboxylates); designed chelation control	DNA crosslinking (anticancer), enzyme inhibition, targeted drug delivery, controlled release
Common Ground	—	Both rely on **metal-ligand interactions** to stabilize complexes and modulate reactivity	Highlight the role of **biocoordination chemistry** in bridging traditional and modern therapeutics

Relevance of Metallodrugs in Ayurveda

Human physiology and biochemistry have reported that alkali, alkaline earth, and transition metals (Inorganic constituents) have around 1-3 percent active presence as structural and functional elements. These inorganic elements contribute centrally by remaining involved in electrolytic balances, cellular and muscular osmosis, cardiovascular regulations, skeletal strength, neurological balances, *etc.* Hence, neither can be overlooked nor marginalized. In Ayurvedic medicine, the therapeutic use of metals is highly sophisticated, involving both biological and chemical transformations that optimize their efficacy and safety. Metals and minerals are purified and transformed through processes such as Shodhana, which removes impurities, and Bhasmikaran, which renders the metals into ash-like forms with altered physicochemical properties. According to Rasa Shastra, an ancient text dedicated to medicinal alchemy, gold (Swarnabhasma) has rejuvenating properties and isused to boost immunity, while silver (Rajatabhasma) is employed for its antimicrobial effects, and copper (Tamrabhasma) is recognized for its benefits in anti-inflammatory action [2]. The processes of purification and calcination create complex structures with highly specific particle sizes and surface properties, leading to enhanced bioavailability and reduced toxicity [3]. The text compiled and edited by Vagbhatta (8th Century AD), provides a lot of wisdom based on innovations in the field of health sciences; this was a time when only a few testing tools were available [4].

Modern analytical studies have confirmed that the traditional methods of Ayurvedic metallurgy result in nanoparticles with unique surface characteristics, supporting both targeted therapeutic action and cellular uptake. For instance, Ayurvedic formulations of gold and silver in bhasma form contain particles of nanometricscale (ultrafine particles) that demonstrate anti-inflammatory and antimicrobial properties, lending scientific credibility to historical uses [5]. These insights suggest that Ayurvedic metallodrugs not only have therapeutic potential but also offer guiding principles for the design of advanced coordination compounds and metallodrugs in modern medicine, where metal-ligand complexes play an essential role in targeted therapies and controlled drug release [6].

The purpose of this chapter is to bridge the wisdom of Ayurvedic metallurgical practices with contemporary developments in coordination chemistry and metallodrug design. This chapter examines the chemical and pharmacological principles behind Ayurvedic metal-based formulations, such as purification, nano-scaling, and bioavailability enhancement, and highlights their relevance in the development of modern metal-based pharmaceuticals (seeking wisdom from ancient Global Knowledge Sources). By comparing traditional approaches to coordination chemistry and drug delivery, this chapter illustrates how ancient methodologies can inspire new approaches to stability, efficacy, and safety in metal-based therapeutics. This ancient knowledge exploration may be helpful to reduce the time of clinical trials of several chemotherapeutics and thus to broaden the domain of alternative medicines on one hand and to reduce the cost of new therapeutics on the other. Ultimately, the chapter covers the value of combining Ayurvedic practices with contemporary scientific advancements, establishing a framework for developing integrative therapeutic strategies that leverage the strengths of both systems for improved health care outcomes.

METALLODRUGS IN AYURVEDA

Metallodrugs, traditionally significant in Ayurvedic medicine, undergo specialized preparative processes that transform their chemical properties, making them safer, more bioavailable, and therapeutically effective. Key processes include Bhasmikaran (calcination) and Shodhana (purification), which not only detoxify metals but also enhance their pharmacological properties. Recent scientific studies have begun to validate and mimic these techniques using modern analytical and nanotechnological methods.

Traditional Preparation Methods

Bhasmikaran (Calcination): Transforming Metals to Bioactive Ash.

Bhasmikaran is a unique Ayurvedic process where metals and minerals are calcined to create a fine, bioactive ash called bhasma. This ash is purported to carry enhanced therapeutic efficacy with minimal toxicity, as the calcination process induces significant chemical and physical transformations.

Chemical Transformation: In Bhasmikaran, repeated heating and cooling cycles facilitate oxidation and phase transitions, converting elemental metals into oxides, sulfides, or other stable compounds, which are generally less reactive and more biocompatible. For instance, bhasma forms are reduced to nano- or submicron particles, enhancing their surface area and interaction with biological systems [1]. This nano-scale particle size has been confirmed by recent spectroscopic and microscopic analyses, linking traditional processing methods with enhanced absorption and bioactivity [7].

Modern Mimicry through Nanotechnology: Researchers have used nanotechnology to mimic Bhasmikaran conditions, aiming to replicate the particle size, stability, and bioactivity of traditional bhasmas. For example, some researchers demonstrated that heating metal nanoparticles under controlled conditions could mimic the outcomes of Bhasmikaran, producing particles with physical characteristics similar to those of traditional gold bhasma. This approach not only validates the traditional technique but also facilitates reproducibility and standardization in metallodrug production.

Shodhana (Purification): Detoxification and Pre-Processing

Shodhana, or purification, is a preliminary step in Ayurvedic metal preparation that removes toxic impurities and alters surface chemistry, preparing metals for safer medicinal use. This process typically involves washing, boiling, or soaking metals in herbal decoctions or acidic solutions, which chemically modify and detoxify the raw materials [8].

Chemical Mechanisms: During Shodhana, metals such as mercury, lead, and copper are treated with herbal extracts or acidic media to remove surface impurities and facilitate partial oxidation. For example, copper is boiled in triphala decoction (a herbal mixture with mildly acidic properties) to remove toxic elements and reduce heavy metal content. Analytical studies using techniques like ICP-MS (Inductively Coupled Plasma Mass Spectrometry) have shown that Shodhana significantly reduces heavy metal concentrations, supporting its detoxifying efficacy.

Enhanced Bioavailability: Shodhana also impacts the particle size and solubility of metals, which can increase their bioavailability. By reducing the size of metal particles and altering surface characteristics, Shodhana enhances the body'sability to absorb and utilize the metallodrugs, thereby increasing their therapeutic potential [9].

Therapeutic Applications of Ayurvedic Metallodrugs

Ayurvedic metallodrugs are traditionally employed in treating a range of chronic conditions, and modern research is uncovering the biochemical basis for these effects. Below are key examples of Ayurvedic metallo drugs, their traditional uses, and scientific validations of their therapeutic properties.

Gold Bhasma (SwarnaBhasma)

Traditional Use: Widely used for its anti-inflammatory and immunomodulatory properties, particularly in managing arthritis, neurological conditions, and boosting immunity.

Scientific Validation: Recent studies show that gold bhasma exhibits anti-inflammatory effects by modulating cytokine release and inhibiting pro-inflammatory pathways. In arthritis models, gold bhasma significantly reduced inflammation and joint damage [10]. The nanoparticle size of gold in bhasma form appears to facilitate its bioactivity by enhancing cellular interactions [11].

Silver Bhasma (Rajata Bhasma)

Traditional Use: Primarily used for its antimicrobial effects to treat infections and wounds.

Scientific Validation: Silver bhasma, prepared through traditional calcination, has shown potent antimicrobial activity, especially against antibiotic-resistant bacteria. Mechanistic studies indicate that silver nanoparticles disrupt bacterial cell membranes, generate Reactive Oxygen Species (ROS), and interfere with microbial DNA, leading to bacterial death [12]. This validates its historical use as a natural antibiotic.

Copper Bhasma (Tamra Bhasma)

Traditional Use: Commonly used to treat liver/digestive and metabolic disorders, including diabetes.

Scientific Validation: Studies have revealed that copper bhasma can aid glucose metabolism and may have hypoglycemic effects. Copper's involvement in key metabolic enzymes, such as cytochrome c oxidase, facilitates cellular respiration and improves insulin sensitivity [13]. This supports copper bhasma's application as an adjunct in diabetes management.

Mercurial Preparations (Rasaushadhi)

Traditional Use: Despite mercury's inherent toxicity, carefully purified mercurial compounds [Hg(II)] are used in small doses for treating skin diseases and infectious ailments, topically.

Scientific Challenges and Validation: Due to safety concerns, mercurial preparations have stringent regulatory limitations. However, studies indicate that traditional Shodhana processes reduce mercury toxicity by removing free elemental mercury and forming a less toxic form, Hg(II) [14]. This highlights the need for controlled usage and further research into its safety profile.

CHELATION CONCEPTS IN AYURVEDIC MEDICINES (HERBOMINERAL FORMULATIONS)

Ayurvedic medicine has long leveraged chelation as a fundamental approach to neutralize both environmental and medicinal heavy metals, harnessing plant-based compounds to bind, deactivate, and facilitate the excretion of toxic metals. It is pertinent to mention that such life-essential metal-chelates and metallic oxides were alternatives to antimicrobial drugs, prior to 1930. Controlled condition use of such metallodrugs is less hepatotoxic and nephrotoxic than several high molecular weight antibiotics. Unlike synthetic chelators such as EDTA or DMSA, which can be harsh on the body and may cause adverse effects by binding even bioessential cations and leading to kidney stress, Ayurvedic chelation utilizes biocompatible herbal compounds that act as natural ligands with high affinity for metals [15]. These agents not only bind metals effectively but also offer antioxidant, anti-inflammatory, and detoxifying benefits, making them uniquely versatile in heavy metal detoxification [16].

Recent scientific studies have provided a clearer picture of the underlying chemical and pharmacological mechanisms driving Ayurvedic chelation. For instance, polyphenolic compounds in Triphala form stable complexes with lead and cadmium ions, demonstrating the potential of these natural ligands to sequester metals and prevent them from interacting harmfully with biological systems. As research continues to decode these traditional practices, new avenues are emerging for integrating Ayurvedic chelation into modern therapeutic frameworks, especially for chronic metal toxicity and related oxidative stress disorders [17]. The Panchakarma principle of Ayurveda is today's detoxification process, which is also nearer to antiaging phenomena.

Understanding Chelation in Ayurveda

Traditional Natural Chelating Agents and Their Chemistry

Ayurveda utilizes a wide range of plant-based chelating agents, including Triphala, Guduchi (*Tinospora cordifolia*), Amalaki (*Phyllanthus emblica*), Haritaki (*Terminalia chebula*), and Neem (*Azadirachta indica*). These herbs contain bioactive compounds, such as polyphenols, flavonoids, and organic acids, that are capable of binding heavy metals, forming stable complexes, and facilitating their excretion from the body [18].

Key Ligands and Chemical Properties in Ayurvedic Herbs

Polyphenols in Triphala: Triphala, a traditional Ayurvedic combination of Amalaki, Haritaki, and Bibhitaki, is rich in chelating agents such as gallic acid, ellagic acid, and tannins. These polyphenols possess multiple hydroxyl groups, which make them particularly effective at complexing metals such as lead and cadmium. A study in Food and Chemical Toxicology demonstrated that Triphala reduced bioavailable lead levels by over 70% in animal models, effectively binding to lead ions and facilitating their excretion [19].

Flavonoids in Guduchi: Guduchi contains quercetin and other flavonoids with hydroxyl and keto functional groups that enable binding with metals such as mercury and cadmium. Studies have shown that quercetin forms stable complexes with mercury through both oxygen and carbonyl groups, enhancing mercury's excretion *via* the urinary tract [20].

Organic Acids and Tannins in Amalaki and Haritaki

Amalaki (Indian gooseberry) is rich in ascorbic acid (vitamin C) and citric acid, both of which chelate iron and copper ions. These acids form stable, water-soluble complexes, reducing the metals' toxicity and promoting renal clearance [21]. Tannins in Haritaki can form multi-dentate complexes with various metals, stabilizing them in less reactive states, which, in turn, helps protect cells from oxidative damage [21].

Breakthrough Studies in Plant-Based Chelation

Neem (*Azadirachta indica*): Neem is a highly versatile Ayurvedic herb recognized for its detoxifying and metal-chelating properties. A study in Environmental Toxicology and Pharmacology demonstrated that neem leaf extract effectively chelated arsenic ions, reducing arsenic accumulation in liver and kidney tissues of animal models. Neem has also been rated as an herbal antibiotic. This study underscored neem's potential in treating arsenic toxicity, which is a serious public health issue in areas with high arsenic contamination [21, 22].

Shilajit: Shilajit, a mineral-rich resin, is a well-known Ayurvedic chelator, particularly for heavy metals like iron and copper. Containing fulvic acid and humic acid, Shilajit binds metals through multiple coordination sites and facilitates their excretion. A study found that Shilajit significantly reduced iron overload in animal models, suggesting its potential use in managing iron toxicity and hemochromatosis (Table **2**) [23].

Table 2. Key ayurvedic ligands and their biocoordination features.

Herb/Source	Active Ligand(s)	Metal Binding Sites	Biological Role	Example Complex
Triphala	Gallic acid, Ellagic acid, Tannins	–OH, –COOH groups	Antioxidant, Metal chelation	Fe–Gallic acid complex
Guduchi	Quercetin, Flavonoids	–OH, –C=O groups	Detox, Anti-inflammatory	Hg–Quercetin complex
Amalaki	Ascorbic acid, Citric acid	–OH, –COOH groups	Iron chelation, Antioxidant	Fe–Ascorbate
Haritaki	Tannins	–OH, –COOH groups	Oxidative stress protection	Cu–Tannin complex
Shilajit	Fulvic acid, Humic acid	–COOH, –OH groups	Metal chelation, Immunomodulation	Fe–Fulvate
Neem	Flavonoids	–OH groups	Antimicrobial, Arsenic chelation	As–Flavonoid complex

Mechanism of Action: Chemical Pathways and Ligand Interactions

The chelationmechanism in Ayurveda relies on the specific ligand properties of natural compounds, which enable them to bind metal ions through coordination bonds, stabilizing the metals and promoting detoxification.

Coordination Chemistry and Stability of Metal-Ligand Complexes

Ayurvedic chelators, particularly polyphenols, form stable coordination complexes with toxic metals by donating electron pairs from hydroxyl, carboxyl, or keto groups. A study revealed that polyphenols in Triphala create stable five-membered rings with metal ions, neutralizing metal toxicity and preventing interaction with biological tissues [24].

Reduction of Bioavailable Metal Ions

Chelation with Ayurvedic ligands reduces the reactivity and bioavailability of toxic metals. For instance, the flavonoid-metal complexes in Guduchi form highly stable bonds that effectively isolate metals, preventing them from participating in redox reactions that can lead to cellular damage [25].

Mechanism of Metal Detoxification

Enhanced Solubility and Excretion: Chelated metal ions typically exhibit higher water solubility, which aids in renal excretion. For example, Guduchi's flavonoids promote mercury or heavy metal excretion by forming water-soluble complexes, making them readily filtered by the kidneys. A study demonstrated that Guduchi treatment in mercury-exposed rats led to a significant reduction in mercury levels in blood and tissues [26].

Inhibition of Oxidative Stress: Free metals like iron and copper catalyze the formation of Reactive Oxygen Species (ROS) *via* Fenton and Haber-Weiss reactions, which are detrimental to cellular integrity. Ayurvedic chelators prevent the reactions by tightly binding to the metals, thereby inhibiting their catalytic activity. Another study demonstrated that the tannins in Haritaki chelate iron ions, inhibiting iron-catalyzed lipid peroxidation and protecting cellular structures [27].

Additional Examples of Metal-Ligand Complexes in Ayurvedic Chelation

Lead-Flavonoid Complexes: The flavonoids in neem leaf extract are known to bind lead ions through oxygen-containing functional groups, reducing lead's neurotoxic impact on neurological tissues. This chelation effect has been shown to decrease lead accumulation in the brain, mitigating neurotoxicity [28].

Arsenic-Sulfur Complexes: Ayurvedic herbs like Ashwagandha contain sulfur compounds that effectively chelate arsenic, which has a high affinity for sulphur (Soft-Soft acid-base principle). A study found that sulfur-rich compounds in Ashwagandha significantly reduced arsenic levels in blood and tissues, providing a potential remedy for arsenic toxicity, especially in regions with contaminated water sources [29].

COORDINATION COMPOUNDS AND MODERN PHARMACEUTICALS

Coordination compounds have revolutionized pharmaceutical science, particularly in cancer treatment and targeted drug delivery. The principles of coordination chemistry allow for precisely engineered metal-ligand complexes, which exhibit tailored reactivity, stability, and biocompatibility for therapeutic applications. Groundbreaking research has paved the way for metallodrugs that not only enhance efficacy but also minimize toxicity, transforming the landscape of modern medicine. This section explores the fundamentals of coordination chemistry and its critical role in the development of metallodrugs and advanced drug delivery systems.

Basics of Coordination Chemistry in Drug Development

Understanding Coordination Compounds and Ligands

Coordination compounds are formed when central metal ions bind to surrounding ligands through coordinate covalent bonds. This metal-ligand interaction creates diverse geometric structures (*e.g.*, octahedral, tetrahedral, or square planar) that are highly influential in determining the compound's chemical reactivity, stability, and biological behaviour. Ligands, the molecules that donate electron pairs to form bonds with the metal, can vary widely in their structure and binding affinity. Monodentate ligands (such as chloride ions or ammonia) bind through a single atom, whereas bidentate and polydentate ligands (like ethylenediamine and EDTA) possess multiple donor atoms, enhancing the stability and therapeutic potential of these complexes [30].

This concept of stability through multiple binding sites, also known as the chelate effect, has been a pivotal principle in developing metallo drugs with predictable pharmacokinetics and reduced side effects. By carefully choosing ligands that can form stable yet bioresponsive metal complexes, researchers can design compounds with specific therapeutic functions, such as delayed release in response to pH changes or selective binding to cancer cells/reactive species, over-expressing certain receptors [31].

Metallodrugs in Modern Medicine

Landmark Drugs and Mechanisms of Action

One of the most significant breakthroughs in modern medicine has been the use of metallodrugs in chemotherapy. Cisplatin, discovered in the 1960s, became a cornerstone of cancer treatment by leveraging its unique platinum-based structure, which enables covalent bonding with purine bases in DNA (Rosenberg *et al.*, 1965). This bonding causes DNA cross-linking that prevents replication and transcription, inducing cell apoptosis. The success of cisplatin spurred the development of other platinum-based compounds, such as carboplatin and oxaliplatin, which introduce modified ligands to reduce toxicity while maintaining anticancer efficacy. Carboplatin, for example, incorporates a bidentate cyclobutane dicarboxylate ligand that slows down the compound's reactivity, lowering nephrotoxicity and making it more suitable for patients with compromised kidney function [32].

Other platinum-based drugs, like oxaliplatin, have been optimized for targeting specific cancers, such as colorectal cancer, through a mechanism that induces less DNA repair and thereby causes more persistent DNA damage in tumor cells. Oxaliplatin's unique DACH (diaminocyclohexane) ligand enhances its antitumor efficacy while reducing some of the peripheral neuropathy common to platinum drugs, marking it as a second-generation platinum compound with substantial clinical benefits [33].

Advances in Metal-Based Cancer Therapies

Recent innovations in metal-based therapies continue to expand the versatility of metallodrugs. For instance, studies have explored the potential of ruthenium-based drugs, such as KP1019 and NAMI-A, which exhibit anticancer properties with significantly lower toxicity than platinum-based compounds. Unlike platinum, ruthenium can mimic iron's behavior in the body, allowing it to circulate through iron transport pathways and to preferentially accumulate in tumors due to increased transferrin receptor expression in cancer cells [34]. This selective accumulation enhances the therapeutic index of ruthenium drugs while minimizing systemic side effects, an attribute highly desirable in chemotherapy [35].

Another study demonstrated the use of platinum-based nanoparticles modified with folic acid to target folate receptors overexpressed on cancer cells. These nanoparticles, a form of "active targeting," allow for the platinum drug to be released precisely where it is needed, sparing healthy tissues and reducing side effects commonly associated with traditional chemotherapy. The use of targeting ligands, such as folic acid, has thus opened new avenues for improving drug delivery specificity, a critical development in reducing off-target effects and improving patient outcomes [36].

Furthermore, recent exploration of gold and copper complexes in cancer therapy has shown promising results. Gold complexes, for instance, leverage the metal's reactivity with sulfur-containing proteins found abundantly in cancer cells, allowing for selective cytotoxicity. The reactivity of these complexes with thiols disrupts cancer cell metabolism without affecting normal cells, thus reducing adverse effects typically associated with chemotherapy [37].

Salvarsan (As-amine complex) was reported in 1908 as a drug for syphilis treatment. Similarly, Pd (II) and Pt(IV) complexes are also under trial for cancer treatment.

Coordination Compounds in Drug Delivery Systems

Targeted Delivery and Controlled Release

Coordination compounds are also instrumental in developing targeted drug delivery systems, owing to their ability to form stable yet responsive complexes that release drugs under specific conditions. For example, coordination-based drug carriers, such as Metal-Organic Frameworks (MOFs), exhibit high porosity, tunable surface properties, and biodegradability, making them ideal for encapsulating and delivering therapeutic agents. These frameworks utilize the principles of coordination chemistry to encapsulate drugs and protect them from premature degradation, ensuring that they are released only when they reach the target site.

A study demonstrated the use of iron-based coordination polymers for selective drug release in the reductive environments of tumors. These polymers were designed to degrade in response to the low oxygen and high glutathione levels characteristic of cancer cells, ensuring a controlled release that spares healthy tissues. This "smart" release mechanism, activated by environmental conditions unique to tumors, highlights the potential of coordination compounds to provide precise therapeutic effects with minimal systemic exposure [38].

Chemical Properties Enabling Selectivity

Coordination compounds offer a versatile platform for controlled release, as the metal-ligand bonds can be tuned to respond to specific physiological triggers such as pH, temperature, or redox states. For instance, platinum-based prodrugs have been designed to remain stable in the bloodstream but activate in the acidic microenvironment of tumors. This selective activation helps target cancer cells more effectively, as demonstrated by researchers [39], who developed a platinum prodrug that releases the active drug upon encountering the acidic pH characteristic of tumor tissues. The study showed a substantial increase in drug efficacy with reduced off-target toxicity, showcasing the importance of tuning metal-ligand chemistry in improving therapeutic precision.

Additionally, metal-ligand complexes can achieve dual functionality by combining therapeutic and diagnostic properties in a single platform, termed "theranostics." An example of this is gadolinium-based coordination complexes, which are widely used in MRI imaging for their paramagnetic properties. Researchers have modified these complexes to incorporate anticancer drugs, enabling real-time tracking of drug distribution while simultaneously delivering therapy. This dual-purpose approach not only enhances drug efficacy but also provides critical data on treatment progress, exemplifying the sophistication possible with a coordination chemist. . The recent world pandemic witnessed frequent use of Zinc salts as an immunobooster; it involves a lot of bioinorganic chemistry and molecular biology. Vanadium has also been reported in past twenty years, as another emerging biometal, due to its fascinating redox- and biocoordination chemistry.

INTEGRATING AYURVEDIC METALLODRUGS WITH MODERN DRUG DELIVERY

Blending Ayurvedic metallurgical principles with modern drug delivery science has opened innovative pathways for developing targeted and biocompatible therapies. Ayurvedic metallic-alchemy employs methods such as Bhasmikaran (calcination) and Shodhana (purification), which transform raw metals into bioactive, stable forms. These principles align closely with modern nano-engineering and coordination chemistry, where nano-scaling, bioavailability, and controlled release are crucial. Here, we indicate only how IKS and Global Knowledge System (GKS) approaches inform the world of the hitherto-unveiled biocoordination chemistry, with a view to ignite future advancements in bioinspired and biopotent coordination compounds, and to examine specific evidence-based case studies that demonstrate successful integration.

Synergy Between Ayurveda and Modern Science

Ayurvedic Metallurgy as a Foundation for Coordination Chemistry

Ayurvedic bhasmas, or calcined metal ash preparations, are renowned for their enhanced bioavailability due to reduced particle size and altered surface characteristics achieved through repeated heating cycles. This ancient practice mirrors current nanotechnological approaches that create metal-based nanoparticles with targeted therapeutic profiles. Research conducted at institutions such as the Indian Institute of Science has demonstrated that these bioactive ligand-stabilized iron nanoparticles mimic traditional Ayurvedic formulations, with marked improvements in antioxidant effects that are promising for managing oxidative stress [40].

Designing Novel Coordination Compounds

Latent Ayurvedic Principles Applied to Modern Coordination Complexes

Nano-scaling, a natural outcome of Bhasmikaran, has proven instrumental in creating modern coordination compounds that can cross biological barriers more effectively. Inspired by this Ayurvedic practice, researchers at Banaras Hindu University (BHU) have synthesized zinc oxide nanoparticles designed to resemble Yashadabhasma (Zinc oxides), with enhanced bioavailability and specific antimicrobial activity. These nanoparticles exhibit therapeutic effects in wound healing that parallel the benefits ascribed to traditional zinc bhasma, validating Ayurvedic concepts in a modern biomedical setting [40].

Moreover, Ayurvedic metallurgical principles help to inform the design of coordination compounds that feature controlled release properties. Metal-Organic Frameworks (MOFs) and coordination complexes can be engineered to respond to environmentalcues, ensuring targeted drug delivery. For example, copper and zinc coordination compounds inspired by bhasma synthesis are engineered with pH-sensitive release properties, allowing for selective drug release within acidic tumor microenvironments. Researchers from the All India Institute of Medical Sciences (AIIMS) have shown that such bioinspired metal-ligand structures deliver drugs more precisely, reflecting Ayurveda's aim of targeting specific tissues.

Case Studies in Bioinspired Coordination Compounds

Examples of Ayurvedic Principles in Coordination Chemistry

The integration of Ayurvedic metallurgical techniques into modern coordination chemistry is gaining traction at key research institutions across India, resulting in notable advancements. Below are examples that illustrate the effective blending of traditional principles with contemporary coordination compound design:

Gold Nanoparticles for Targeted Cancer Therapy

Researchers have developed gold nanoparticles inspired by the Ayurvedic gold bhasma. These nanoparticles were engineered with surface properties tailored to increase cellular uptake by cancer cells, exerting selective cytotoxicity through apoptosis induction while minimizing effects on healthy cells. This research highlights how Ayurvedic approaches to gold bhasma synthesis can enhance precision in modern cancer treatments [41].

Iron-Based Antioxidant Therapy from Ayurvedic Inspiration

Researchers have synthesized iron coordination complexes using stabilizing ligands derived from Ayurvedic herbs such as Triphala. These iron-based compounds mimic the antioxidant activities of traditional Ayurvedic bhasmas, successfully neutralizing Reactive Oxygen Species (ROS) implicated in neurodegenerative diseases. This study underscores the synergy between Ayurvedic herbal extracts and coordination chemistry in developing therapeutic agents that manage oxidative stress [42].

Copper Complexes for Antimicrobial Applications

Inspired by the Ayurvedic copper preparation known as Tamra bhasma, researchers developed copper complexes using herbal ligands that replicate the organic environment of traditional bhasmas. These bioinspired copper complexes demonstrated potent antimicrobial effects, especially against drug-resistant bacterial strains, underscoring their relevance in developing novel antimicrobial agents. This work highlights the applicability of Tamra bhasma's antimicrobial principles in combating modern antibiotic resistance.

Zinc Oxide Nanoparticles for Enhanced Wound Healing

Zinc oxide nanoparticles designed to emulate Yashada bhasma have shown efficacy in wound healing applications. A study applied Ayurvedic insights to create bioactive zinc nanoparticles that accelerate wound healing and exhibit antimicrobial properties. This bioinspired approach validates Ayurvedic applications of zinc bhasma in dermatology and aligns with advanced therapeutic needs for chronic wound management [42].

CHALLENGES AND CONSIDERATIONS IN INTEGRATING AYURVEDIC METALLODRUGS WITH MODERN PHARMACEUTICAL STANDARDS

Integrating Ayurvedic metallodrugs into modern pharmaceutical systems presents distinct challenges, particularly in the areas of standardization, safety, and regulatory compliance. While Ayurvedic formulations offer significant therapeutic potential, issues such as variability in metal content, particle size, and potential toxicity necessitate rigorous quality control and safety assessments.

Additionally, regulatory frameworks of ten present hurdles due to the complex nature of traditional metallodrugs, which blend both biological and metallic elements.

FUTURE PERSPECTIVES

The integration of Ayurvedic metallodrugs with modern pharmaceutical science presents transformative possibilities for advancing health care globally. Future research in this field highlights exciting interdisciplinary opportunities to combine traditional Ayurvedic principles with modern drug development, particularly in chelation therapies, nano-scaled metal drugs, and bioinspired drug delivery systems. The biocoordination chemistry coupled with therapeutic efficacy is also associated with stable- and metastable metallobiomolecules; their formation and deformation in physiological conditions in real time, and their geometry and structural biocompability during biotransportation and transformation, are other challenging issues to be reviewed. Molecular biology, molecular-chemical dynamics and reactivity, and electro-oxidative changes also attract scientific discoveries in the coming decades of personalised medicines.

CONCLUSION

This chapter has explored the intricate connections and emerging demand between Ayurvedic metallurgical practices- an experiential knowledge *vs.* experimental knowledge, and modern coordination chemistry, chelation science, and metallodrug developments. By examining traditional Ayurvedic processes such as Bhasmikaran and Shodhana, we uncover chemical principles like nano-scaling, purification, and complexation, which serve as a foundation for contemporary approaches in coordination chemistry. These processes are not only pivotal to creating bioavailable and stable Ayurvedic bhasmas but also directly inform the design of modern metallodrugs, where particle size, metal-ligand stability, and controlled release are essential to tailor the therapeutic efficacy.

The ancient Ayurvedic techniques aim to unveil the novel metal-ligand interactions that enhance bioavailability, minimize toxicity, and enable precise therapeutic targeting. By integrating metals with organic and biocompatible ligands, researchers can craft coordination compounds with properties akin to those observed in Ayurvedic metallodrugs, such as improved absorption, cellular uptake, and selectivity. These principles support the development of cutting-edge metallodrugs that target cancer cells, infectious agents, and inflammation, leveraging the complexation and chelation approaches embedded in Ayurvedic wisdom.

Chelation, a core component in Ayurvedic detoxification and rejuvenation therapies, has also been highlighted as a bridge between traditional and modern methods. Ayurvedic chelators,often derived from herbal ligands, bind metals to mitigate toxicity and facilitate excretion, a process that mirrors contemporary detoxification therapies in pharmacology. Advanced analytical techniques such as ICP-MS, AAS, TEM, Electrochemical analyser, and spectroscopy have validated the efficacy of Ayurvedic chelation, offering insights into how traditional chelating agents can provide information about safer, more effective detoxification approaches in modern medicine.

Looking toward the future, of the integration of Ayurvedic principles with bio-coordination chemistry (science deals in a focused way on metallo-biomolecules and metallodrugs), it offers a transformative approach to future drug design and therapeutic development strategy, in search of an alternative medicinal system. This interdisciplinary model encourages collaborations among chemists, pharmacologists, toxicologists, and traditional practitioners to deepen our understanding of bioinspired metal-based therapies. By harmonizing these fields, we can unlock innovative metallodrugs that are tailored for enhanced bioavailability, safety, and efficacy, while remaining aligned with traditional Ayurvedic goals of individualized, balanced treatment. Ultimately, this holistic approach has the potential to impact global healthcare by providing culturally resonant and scientifically rigorous therapies that honour both ancient knowledge and modern innovation, in this area.

REFERENCES

[1] Pal D, Gurjar VK. Nanometals in bhasma: Ayurvedic medicine. In: Rai M, Shegokar R, Eds. Metal Nanoparticles in Pharma. Cham: Springer International Publishing 2017; pp. 389-415. [http://dx.doi.org/10.1007/978-3-319-63790-7_17]

[2] Goswami AK, Kostova I. Medicinal and biological inorganic chemistry. Berlin: Walter de Gruyter GmbH & Co KG 2022. [http://dx.doi.org/10.1515/9781501516115]

[3] Shukla S, Mishra AP. Medicinal importance of inorganic materials. In: Srivastava A, Awasthi K, Tripathi M, Eds. Versatile Solicitations of Materials Science in Diverse Science Fields. New York: Nova Science Publishers 2021.

[4] Ray P. Chemistry in ancient India. J Chem Educ 1948; 25(6): 327-31. [http://dx.doi.org/10.1021/ed025p327]

[5] Lokhande JN, Lokhande S, Mahajan S, Shinde G, Pathak Y. Botanical drug products of ayurved, indian system of medicine in integration with other metallopharmaceuticals, nano medical devices, and theranostics. Botanical Drug Products. Boca Raton: CRC Press 2018; pp. 139-58. [http://dx.doi.org/10.1201/9781315369273]

[6] Riccardi L, Genna V, De Vivo M. Metal–ligand interactions in drug design. Nat Rev Chem 2018; 2(7): 100-12. [http://dx.doi.org/10.1038/s41570-018-0018-6]

[7] Sreelakshmi S, Vineeth PK, Mohanan A, Ramesh NV. Ayurvedic bhasma and synthesized

nanoparticles: A comparative review. Mater Today Proc 2021; 46: 3079-83. [http://dx.doi.org/10.1016/j.matpr.2021.02.585]

[8] Sonar SPJ. A comparative pharmaceutico-analytical study of vanga bhasma prepared by three purificatory methods: Samanya shodhana, vishesha shodhana, samanya & visesha shodhana. Bangalore: Rajiv Gandhi University of Health Sciences 2014.http://rguhs.ac PhD Thesis

[9] Raisuddin S. Ayurvedic bhasmas. Scientific basis for ayurvedic therapies. London: Routledge 2003; pp. 107-24. [http://dx.doi.org/10.1201/9780203498583.ch6]

[10] Rani S, Magluria A, Sharma U, Mitra S, Sharma KC. Origin, occurrence, pharmaceutical, analytical, and various pharmacological studies on Swarna bhasma (~incinerated gold ash): A synoptic review. Int J Ayurveda Res 2024; 5(2): 76-84. [http://dx.doi.org/10.4103/ijar.ijar_170_23]

[11] Netam , Bhargava V, Singh R, Sharma P. Physico-chemical characterization of ayurvedic swarna bhasma by using SEM, EDAX, XRD and PSA. Journal of Complementary Medicine Research 2021; 12(2): 204-9. [http://dx.doi.org/10.5455/jcmr.2021.12.02.23]

[12] Sinha K, Chaudhury SS, Ruidas B, *et al.* Role of modern biological techniques in evidence-based validation of ayurvedic herbometallic preparations. In: Mandal SC, Chakraborty R, Sen S, Eds. Evidence based validation of traditional medicines. Singapore: Springer 2021; pp. 313-36. [http://dx.doi.org/10.1007/978-981-15-8127-4_16]

[13] Bafna PS, Patil SD. Physicochemical characterisation and anti-inflammatory activity of ayurvedic herbo-metallic *Tamra bhasma* in acute and chronic models of inflammation. Mater Technol 2018; 33(10): 681-8. [http://dx.doi.org/10.1080/10667857.2018.1494241]

[14] Joshi N, Dash MK, Panda P. Critical review on the concept of kajjali: The boon of ayurvedic herbomineral preparations (Rasaushadhi). International Journal of Green Pharmacy 2018; 11: 112-5.

[15] Chaudhary A, Singh N. Herbo mineral formulations (rasaoushadhies) of ayurveda an amazing inheritance of ayurvedic pharmaceutics. Anc Sci Life 2010; 30(1): 18-26. [http://dx.doi.org/10.4103/0257-7941.90223] [PMID: 22557419]

[16] Prajapati PK, Mitra S, Shukla VJ, Ravishankar B. Impact of Bhavana Samskara on physico-chemical parameters with special reference to Gandhaka Rasayana prepared by different media and methods. Ayu 2010; 31(3): 382-6. [http://dx.doi.org/10.4103/0974-8520.77155] [PMID: 22131744]

[17] Sharma R, Amin H, Galib R, Prajapati PK. Molecular targets of common Ayurvedic herbal antioxidants. Journal of Ayurvedic and Herbal Medicine 2017; 3(1): 33-7. [http://dx.doi.org/10.31254/jahm.2017.3106]

[18] Gautam VP, Gautam SP, Ranawat RS, Jain PK. Rationality behind the use of metals and minerals in ayurveda. Int J Pharm Biol Arch 2014; 5(4): 26-9.

[19] Tiwari R, Latheef SK, Ahmed I, *et al.* Herbal immunomodulators - A remedial panacea for designing and developing effective drugs and medicines: Current scenario and future prospects. Curr Drug Metab 2018; 19(3): 264-301. [http://dx.doi.org/10.2174/1389200219666180129125436] [PMID: 29380694]

[20] Dhama K, Sachan S, Khandia R, *et al.* Medicinal and beneficial health applications of Tinospora cordifolia (Guduchi): A miraculous herb countering various diseases/disorders and its immunomodulatory effects. Recent Pat Endocr Metab Immune Drug Discov 2017; 10(2): 96-111. [http://dx.doi.org/10.2174/1872214811666170301105101] [PMID: 28260522]

[21] Upoma SSK. Investigation of *in vitro* antioxidant and cytotoxic potential of triphala methanol extract in different ratios. Dhaka: Brac University 2020. PhD Thesis.

https://dspace.bracu.ac.bd/xmlui/bitstream/ handle/10361/15510/16146036_PHR.pdf?sequence=1

[22] Seriana I, Akmal M, Darusman D, Wahyuni S, Khairan K, Sugito S. Neem leaf (*Azadirachta indica* A. Juss) ethanolic extract on the liver and kidney function of rats. ScientificWorldJournal 2021; 2021: 1-7.
[http://dx.doi.org/10.1155/2021/7970424] [PMID: 33859543]

[23] Bhavsar SK, Thaker AM, Malik JK. Shilajit. In: Gupta RC, Ed. Nutraceuticals. London: Elsevier 2016; pp. 707-16.
[http://dx.doi.org/10.1016/B978-0-12-802147-7.00051-6]

[24] Tarasiuk A, Mosińska P, Fichna J. Triphala: current applications and new perspectives on the treatment of functional gastrointestinal disorders. Chin Med 2018; 13(1): 39.
[http://dx.doi.org/10.1186/s13020-018-0197-6] [PMID: 30034512]

[25] Flora SJS, Pachauri V. Chelation in metal intoxication. Int J Environ Res Public Health 2010; 7(7): 2745-88.
[http://dx.doi.org/10.3390/ijerph7072745] [PMID: 20717537]

[26] Mirkov I, Stojković D, Aleksandrov AP, *et al.* Plant extracts and isolated compounds reduce parameters of oxidative stress induced by heavy metals: An up-to-date review on animal studies. Curr Pharm Des 2020; 26(16): 1799-815.
[http://dx.doi.org/10.2174/1381612826666200407163408] [PMID: 32264808]

[27] Tirgar PR. Investigation into mechanism of action and pharmacological evaluation of various extracts of triticum aestivum (wheat) grass with special reference to its beneficial effects on iron-overload and blood disorders. Rajkot: Saurashtra University 2011.https://core.ac.uk/download/pdf/11822154.pdf PhD Thesis

[28] Virgolini MB, Aschner M. Molecular mechanisms of lead neurotoxicity. In: Aschner M, Costa LG, Eds. Advances in Neurotoxicology. Cambridge, MA: Elsevier 2021; Vol. 5: pp. 159-213.

[29] Alamgir A N M. Herbal drugs: Their collection, preservation, and preparation; evaluation, quality control, and standardization of herbal drugs. Therapeutic Use of Medicinal Plants and Their Extracts: Volume 1. Cham: Springer 2017; 73: pp. Progress in Drug Research453-95.
[http://dx.doi.org/10.1007/978-3-319-63862-1_10]

[30] Atkins PW, Friedman RS. Molecular Quantum Mechanics. 5th ed., Oxford: Oxford University Press 2011.

[31] Frezza M, Hindo S, Chen D, *et al.* Novel metals and metal complexes as platforms for cancer therapy. Curr Pharm Des 2010; 16(16): 1813-25.
[http://dx.doi.org/10.2174/138161210791209009] [PMID: 20337575]

[32] Kenny RG, Marmion CJ. Toward multi-targeted platinum and ruthenium drugs—A new paradigm in cancer drug treatment regimens? Chem Rev 2019; 119(2): 1058-137.
[http://dx.doi.org/10.1021/acs.chemrev.8b00271] [PMID: 30640441]

[33] Trecarichi A. Investigation into Mitochondrial DNA, mitochondrial function and nerve activity in oxaliplatin-induced peripheral neuropathy. London: King's College London 2021. https://kclpure.kcl.ac.uk/portal/files/159976024/2021_Trecarichi_Annalisa_1523917_ethesis_PUREVersion.pdf PhD Thesis

[34] Levina A, Chetcuti ARM, Lay PA. Controversial role of transferrin in the transport of ruthenium anticancer drugs. Biomolecules 2022; 12(9): 1319.
[http://dx.doi.org/10.3390/biom12091319] [PMID: 36139158]

[35] Han Ang W, Dyson PJ. Classical and non-classical ruthenium-based anticancer drugs: towards targeted chemotherapy. Eur J Inorg Chem 2006; 2006(20): 4003-18.
[http://dx.doi.org/10.1002/ejic.200600723]

[36] Chehelgerdi M, Chehelgerdi M, Allela OQB, *et al.* Progressing nanotechnology to improve targeted cancer treatment: overcoming hurdles in its clinical implementation. Mol Cancer 2023; 22(1): 169.

[http://dx.doi.org/10.1186/s12943-023-01865-0] [PMID: 37814270]

[37] Zhao Y, Butler EB, Tan M. Targeting cellular metabolism to improve cancer therapeutics. Cell Death Dis 2013; 4(3): e532.
[http://dx.doi.org/10.1038/cddis.2013.60] [PMID: 23470539]

[38] Suárez-García S, Solórzano R, Alibés R, Busqué F, Novio F, Ruiz-Molina D. Antitumour activity of coordination polymer nanoparticles. Coord Chem Rev 2021; 441: 213977.
[http://dx.doi.org/10.1016/j.ccr.2021.213977]

[39] Zhang C, Xu C, Gao X, Yao Q. Platinum-based drugs for cancer therapy and anti-tumor strategies. Theranostics 2022; 12(5): 2115-32.
[http://dx.doi.org/10.7150/thno.69424] [PMID: 35265202]

[40] Sharma R, Prajapati PK. Nanotechnology in medicine: Leads from Ayurveda. J Pharm Bioallied Sci 2016; 8(1): 80-1.
[http://dx.doi.org/10.4103/0975-7406.171730] [PMID: 26957877]

[41] Sharma CP, Paul W. Blood compatibility studies of Swarna bhasma (gold bhasma), an Ayurvedic drug. Int J Ayurveda Res 2011; 2(1): 14-22.
[http://dx.doi.org/10.4103/0974-7788.83183] [PMID: 21897638]

[42] Pathak PC, Gadgoli CH. Exploring the efficacy of panchavalkal extract and Zinc-Copper Bhasma in promoting wound healing in incision and excision wound models in the rat. J Ethnopharmacol 2024; 320: 117404.
[http://dx.doi.org/10.1016/j.jep.2023.117404] [PMID: 37979819]

CHAPTER 8

Rhenium Coordination Compounds as Promising Radiopharmaceuticals

Priya Yadav[1], **Noopur Srivastava**[2,*] and **Nisha Saxena**[3]

[1] *Department of Chemistry, Amity Institute of Applied Sciences, Amity University, Noida, Uttar Pradesh, India*

[2] *Department of Chemistry & Biochemistry, Sharda School of Engineering & Sciences (SSES), Sharda University, Greater Noida, Uttar Pradesh, India*

[3] *Department of Chemistry, M.R.M. College (Lalit Narayan Mithila University), Darbhanga, Bihar, India*

Abstract: Radiopharmaceuticals are compounds with radioactive isotopes that are used in nuclear medicine to diagnose and treat patients. Rhenium compounds have attracted considerable interest in bioinorganic chemistry, particularly for their potential applications in cancer therapy. Rhenium, a rare transition metal, is noted for its distinct oxidation states and diverse coordination chemistry, which enable the creation of a wide range of complexes. The theragnostic potential of rhenium coordination compounds stems from their capacity to be recognized with different isotopes, allowing for imaging and therapeutic uses. Due to their versatility, rhenium coordination compounds have been studied as radiopharmaceuticals for diagnostic and therapeutic applications. Rhenium-based imaging agents, notably those tagged with the radioactive isotope ^{188}Re, have shown promise in a variety of nuclear imaging techniques, including single-photon emission computed tomography and positron emission tomography. Rhenium(I) complexes are a promising family of anticancer medicines that have various advantages over traditional platinum-based treatments such as cisplatin. Their unique modes of action, enhanced potency in specific cancer cell lines, and ability to overcome drug resistance make them attractive candidates in cancer therapy. Furthermore, rhenium compounds may have a more favorable side effect profile and improved biodistribution, resulting in higher patient tolerance. Rhenium compounds have promise in the creation of targeted cancer medicines. Their distinct chemical features and modes of action provide substantial advantages over standard chemotherapeutics, particularly in terms of cancer cell selectivity and decreased side effects. Rhenium complexes' propensity to cause apoptosis and other forms of programmed cell death, together with their potential for combination therapy, makes them promising prospects in the fight against cancer.

* **Corresponding author Noopur Srivastava:** Department of Chemistry & Biochemistry, Sharda School of Engineering & Sciences (SSES), Sharda University, Greater Noida, Uttar Pradesh, India; E-mail: noopursriv@gmail.com

Abhay Nanda Srivastva (Ed.)
All rights reserved-© 2026 Bentham Science Publishers

Keywords: Applications, Chemical properties, Nuclear medicine, Rhenium compounds, Radiopharmaceuticals, Radioisotopes, Scheme, Therapeutic agent, Theragnostic potential.

INTRODUCTION

Radiopharmaceuticals are radioactive isotope-containing substances used in nuclear medicine for diagnosis and treatment. These substances are delivered into the body and accumulate in certain organs or tissues, enabling the viewing and quantification of physiological processes. These compounds emit gamma rays that are detected by specialized cameras, producing images of their distribution throughout the body. In therapeutic applications, the radiation generated by the radiopharmaceutical is utilized to target and kill sick cells. Overall, radiopharmaceuticals serve an important role in nuclear medicine by allowing for non-invasive imaging and focused treatment of a variety of illnesses [1]. Of the various radionuclides examined, compounds containing rhenium have garnered significant interest due to their outstanding nuclear properties and possibilities for targeted delivery [2]. Isotopes of rhenium, including ^{186}Re and ^{188}Re, possess suitable half-lives, emission energies, and biodistribution properties, which render them attractive options for use in radiopharmaceuticals [2, 3]. Rhenium is capable of forming stable complexes with various ligands, encompassing both organometallic and coordination compounds. These complexes can be engineered to specifically target particular biological processes or receptors, thereby enhancing their selectivity and effectiveness against tumors. The adjustable chemical structures of rhenium compounds enable the creation of personalized radiopharmaceuticals that may improve treatment outcomes. A major advantage of rhenium coordination compounds is their adaptability for tailored clinical applications. By incorporating rhenium into different ligand systems, researchers can develop radiopharmaceuticals that possess improved targeting efficiency, enhanced pharmacokinetics, and reduced toxicity levels [4]. The range of rhenium coordination chemistry has led to the development of numerous rhenium-based radiopharmaceuticals, each possessing unique characteristics and uses. Rhenium-based radiopharmaceuticals provide several benefits compared to other radionuclides utilized in nuclear medicine [1]. The stability of the rhenium-ligand bond is essential for ensuring that the radiopharmaceutical stays in the body until it arrives at the target area. For instance, adjusting temperature and pH can optimize labeling yield, as seen in the synthesis of ^{186}Re-DTPA complexes, which necessitate specific conditions for effective binding. Rhenium complexes are also being explored for their potential in medicinal uses, particularly in targeted radiotherapy. The beta particles from rhenium isotopes can effectively irradiate tumors, providing a focused therapeutic option while minimizing harm to surrounding healthy tissues. The ability to focus on specific cells with rhenium-labeled antibodies enhances the treatment's efficacy [5]. Rhenium compounds can be attached to targeting agents like antibodies or peptides, enabling them to specifically attach to cancerous cells. This customized approach facilitates direct radiation delivery to the tumor while reducing harm to surrounding healthy tissues. The use of bifunctional ligands simplifies the connection of rhenium to these targeted therapies, enhancing the specificity of the treatment [5]. Rhenium compounds can be combined with technetium to create dual-imaging probes. These probes utilize the high sensitivity of nuclear imaging techniques, like SPECT, provided by technetium, alongside the photoluminescent properties of rhenium, enhancing imaging capabilities. For instance, Re-Tc conjugates have been engineered to merge the benefits of both imaging techniques, delivering better spatial resolution and sensitivity.

By allowing focused radiation delivery straight to the tumor, ^{188}Re–HDD-lipiodol reduces the dose to surrounding healthy tissues. This focused approach can improve therapeutic efficacy while minimizing potential adverse effects [6], (Fig. **1**). ^{188}Re-HEDP (Fig. **5**), also known as ^{188}Re-mecasulfate (1-hydroxyethyliden--1,1-diphosphonic acid), is predominantly used as a radiopharmaceutical to treat bone discomfort caused by metastatic cancer (Fig. **1**) [7].

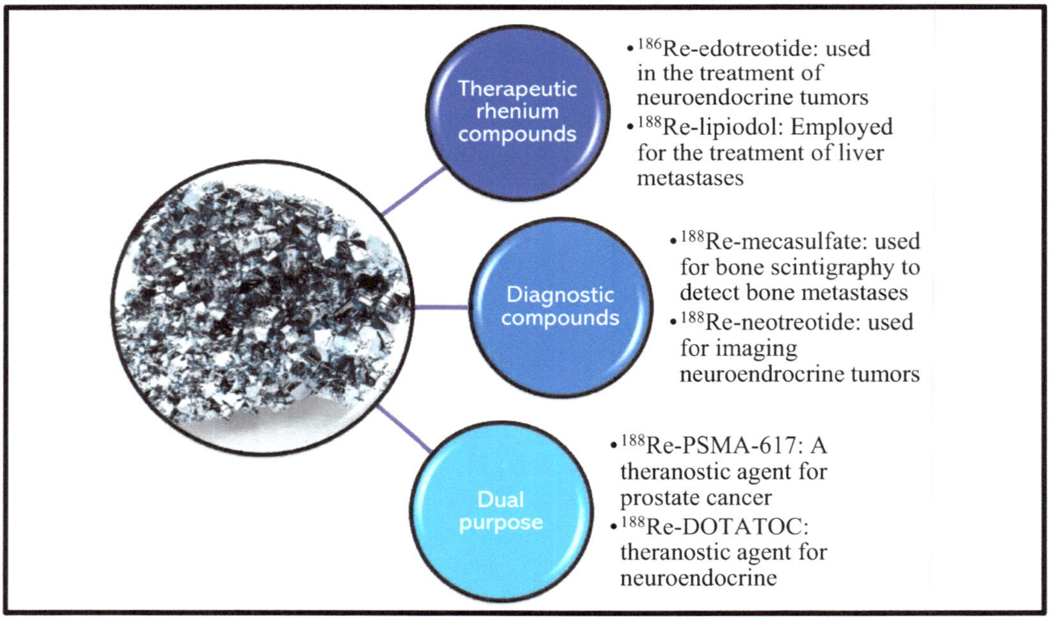

Fig. (1). Different types of rhenium compounds used in radiopharmaceutics.

Chemistry of Rhenium

Walter Noddack, Ida Tacke, and Otto Berg discovered rhenium, which initiated extensive research into its coordination chemistry. The binary chemical Re_2O_7 was produced through direct contact of heated rhenium metal and oxygen. This chemical was notable since it was one of the first oxides of rhenium to be identified. Oxides and oxo-complexes were studied extensively, with rhenium oxides defined and their structures explored using a variety of techniques, including Raman spectroscopy. Throughout the 1960s and 1970s, there was a notable increase in the number of rhenium-related papers, indicating a growing interest in its coordination chemistry and complex synthesis. In the 1980s, investigations started to focus on the applications of rhenium in catalysis, biological chemistry, and materials science, marking a shift from purely synthetic studies to more practical research [8]. Rhenium possesses a broad spectrum of oxidation states, enhancing its adaptability in coordination chemistry and various fields such as catalysis and medicinal chemistry. The oxidation states of rhenium that are most commonly encountered are +1, +2, +4, +5, +6, and +7. Depending on its oxidation state and the nature of the ligands, rhenium shows coordination numbers that range from 4 to 9. Typical coordination numbers include 4 (tetrahedral), 5 (trigonal bipyramidal), and 6 (octahedral). Rhenium can create complexes with a diverse array of ligands, which encompass both bidentate and polydentate ligands. This characteristic is particularly advantageous in the development of radiopharmaceuticals, which necessitate high stability and selectivity. Ligands such as phosphines, amines, and carboxylates can serve as chelating agents, forming stable rhenium chelate rings. This enhances the stability of the complexes and may improve their biological properties [9]. In the +7 oxidation state, rhenium can create complexes with a wider range of ligands, such as oxo and halide ligands, as well as chelating agents like NOTA and NODAGA, which are frequently utilized in radiopharmaceuticals. Rhenium is capable of forming tricarbonyl compounds that typically have the structure fac-[M(CO)3(L)], where L represents a ligand. These compounds exhibit excellent *in vivo* stability and pharmacokinetics, making them suitable for use as SPECT imaging agents. The inclusion of CO ligands enhances the stability and solubility of the complexes, which is beneficial for their application in biological systems [10]. Rhenium isotopes, especially ^{186}Re and ^{188}Re, have important applications in nuclear medicine due to their attractive decay properties, production methods, and specific activities. Their capacity to deliver focused radiation makes them useful in both therapeutic and diagnostic applications in oncology [9]. Table **1** shows some properties of rhenium isotopes.

Table 1. Properties of rhenium isotopes.

Rhenium Isotopes	Production Method	Decay Properties	Specific Activity	Applications in Medicine
Rhenium-186	^{186}Re can be created by irradiating ^{185}Re with neutrons in a nuclear reactor using the neutron capture reaction: $^{185}Re(n,\gamma)^{186}Re$	^{186}Re has a half-life of around 89.2 hours. The primary decay mode is beta emission (β-) to stable ^{186}Os (osmium). Energy: The beta particles emitted have an average energy of around 1.075 MeV.	The specific activity of ^{186}Re varies depending on the production process, although it is typically lower than that of ^{188}Re due to its longer half-life and the nature of its manufacture.	It is primarily used in radiotherapy to treat bone metastases, as well as radiolabeled chemicals for imaging and therapy.
Rhenium-188	^{188}Re is commonly obtained by an ^{188}W/^{188}Re generator, which extracts ^{188}Re from the decay of ^{188}W. The degradation processgoes as follows: $^{188}W \rightarrow ^{188}Re$ + beta particles	Half-life: ^{188}Re has a shorter half-life of approximately 17.0 hours. The decay mode is beta emission (β-), resulting in stable ^{188}Os (osmium). Energy: The beta particles emitted have a greater energy of around 2.11 Mev.	^{188}Re has a higher specific activity than ^{186}Re, making it ideal for therapeutic applications such as bone pain palliation and targeted cancer treatment.	Widely utilized in targeted radiation, particularly for treating disorders like bone pain from metastatic cancer, and in SPECT imaging because of its good decay properties and the ability to form stable complexes with diverse ligands.

Rhenium^{188} and Rhenium^{186} Radiopharmaceuticals

Rhenium-188 is a β-emitting radionuclide with a short half-life of 16.9 hours, making it an ideal choice for cancer treatment. Recent investigations have shown that ^{188}Re-labeled therapeutic radiopharmaceuticals can be successfully used in several early-phase clinical trials to treat a variety of primary cancers, bone metastases, rheumatoid arthritis, and endocoronary therapies. The availability of chemical methods that enable the production of ^{188}Re radiopharmaceuticals in high yield and under physiological settings has created a new, appealing opportunity to use this radionuclide for medicinal purposes. The rise in malignant tumours has accelerated the development of new clinical procedures for cancer treatment, including radionuclide-based therapies utilized in nuclear medicine [9]. Rhenium-186 is a β-emitting radionuclide with a half-life of 3.7 days, making it ideal for therapeutic purposes. ^{186}Re is a popular choice because of its long half-life and ability to emit a γ ray with an energy similar to that of ^{99m}Tc. The biodistributions of ^{186}Re radiopharmaceuticals can be tracked using the same γ camera instrumentation as ^{99m}Tc agents. Rhenium-based radiopharmaceuticals have been successfully used in a variety of clinical trials, demonstrating their potential in nuclear medicine and cancer treatment [9, 11 - 13].

Theragnostic Potential of Rhenium Coordination Compounds

The theragnostic potential of rhenium coordination compounds arises from their ability to be labeled with various isotopes, which enables both imaging and therapeutic applications. For example, the same rhenium-based molecule can be tagged with rhenium-186 for therapeutic use and rhenium-188 for diagnostic imaging, allowing for more individualized treatment planning and monitoring. Furthermore, the flexibility of rhenium coordination chemistry enables the development of targeted radiopharmaceuticals, in which the rhenium complex is conjugated to a specific targeting moiety, such as a peptide or a small molecule, to facilitate the selective delivery of the radionuclide to the desired target [14 - 16]. Imagine a drug that could diagnose and treat an illness at the same time. That is the promise of theragnostic drugs, and rhenium coordination compounds are emerging as strong contenders. Rhenium, a versatile element, has a unique set of qualities that make it perfect for dual-purpose treatments. ^{186}Re, with its prolonged lifespan, functions as a stealthy warrior, delivering a high dosage of radiation to tumours while avoiding injuring healthy cells. ^{188}Re, on the other hand, functions as a beacon, producing gamma rays that allow doctors to precisely locate and evaluate the size of tumours. Scientists can develop theragnostic medicines that recognize and attack diseases by combining these two isotopes in a single molecule [17].

Synthesis of Rhenium Complexes

The synthesis and characterisation of rhenium complexes are crucial stages in the development of rhenium-based radiopharmaceuticals. These techniques guarantee that the complexes possess the necessary traits for effective application in nuclear medicine. Generally, the synthesis begins with precursors of rhenium, such as rhenium oxides or rhenium halides, which can engage with several types of ligands. These ligands may include bidentate and multidentate chelators that help stabilize the rhenium core. Factors such as temperature, choice of solvent, and duration of the reaction are all important considerations for the synthesis process. For instance, reactions can be conducted in organic solvents under inert conditions to prevent the oxidation or hydrolysis of rhenium species. The selection of ligands is essential because it affects the oxidation state of rhenium and the overall stability of the complex. Ligands can be engineered to improve the pharmacokinetic properties and targeting abilities of the resulting radiopharmaceuticals. Rhenium can exist in multiple oxidation states (+1, +2, +4, +5, +7). The targeted oxidation state of the complex is often achieved through precise management of the reaction conditions and careful ligand selection [18]. Multidentate ligands have the ability to form multiple bonds with a single metal ion by utilizing two or more donor atoms. Tetradentate ligands are defined as those that can bond with a metal ion through four donor atoms. These ligands can contain two nitrogen atoms and two sulfur atoms, enabling them to create stable complexes with metal ions like technetium and rhenium (Fig. **2**) [19]. Sulfur donor ligands are a category of ligands where sulfur atoms can bond with metal centers in coordination complexes. There are proposals to pair rhenium with donor atoms like oxygen, sulfur, and even selenium. This highlights a keen interest in exploring the properties and possible uses of rhenium complexes featuring sulfur donor ligands (Fig. **2**) [20].

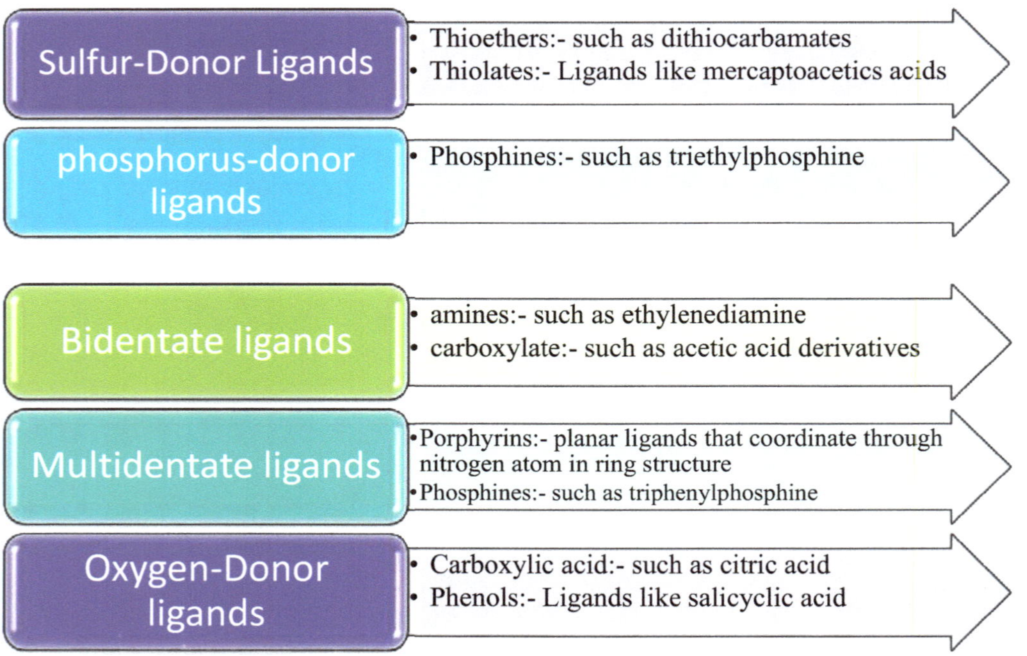

Fig. (2). Different types of ligands used in rhenium complexes.

The rhenium(III) complexes $ReCl_2(pimH)(PPh_3)_2$ and $ReCl_3(pimMe)(PPh_3)$ were produced by reacting $[ReOCl_3(PPh_3)_2]$ with one equivalent of the ligands 2-(2'-pyridyl)imidazole (pimH) and 2-(2'-pyridyl)-1-methylimidazole (pimMe) in ethanol, under reflux. After 1 hour of heating, the reaction resulted in a color change from lime-green to dark purple, and finally to orange precipitates (Scheme **1**) [21].

$$ReOCl_3(PPh_3)_2 + pimH \rightarrow ReCl_2(pimH)(PPh_3)_2 + H_2O$$

$$ReOCl_3(PPh_3)_2 + pimMe \rightarrow ReCl_3(pimMe)(PPh_3) + H_2O$$

Scheme 1. Synthetic route of Re(III) complexes.

The production of a novel metallocarboxylic acid, CpRe(CO)(NO)COOH, resulting from the treatment of $CpRe(CO)_2(NO)^+BF_4^-$ with aqueous KOH. This reaction produces an orange powder that can be identified *via* spectral and elemental analysis. In the presence of Na_2CO_3, the metallocarboxylic acid interacts with Ph_3SnCl and $Re(CO)_4(PPh_3)(F\text{-}BF_3)$ to generate two separate CO_2-bridged complexes: $CpRe(CO)(NO)(CO_2)SnPh_3$ (complex 3) and $CpRe(CO)(NO)(CO_2)Re(CO)_3(PPh_3)$ (complex 4) (Scheme **2**). These compounds show several types of p2-q3 bonding between the bridging CO_2 ligands. X-ray crystallography was used to describe both complexes, and it confirmed the CO_2 ligand's p2-q3 binding. The structural changes between the complexes are related to their IR and ^{13}C NMR spectrum characteristics, revealing variations in bonding and coordination modes. The variances in DRIFTS data for the complexes indicate that the CO_2 ligand's p2-q3 coordination produces separate spectrum signatures. For example, complex 3 has a bigger difference in vibrational frequencies than complex 4, indicating unsymmetrical and symmetrical bonding properties, respectively. The work underlines the ability of these rhenium complexes to activate CO_2, which is important for chemical synthesis and the development of new catalytic processes. The structural and spectral data provide insights into how different bonding conditions can affect the reactivity of CO_2 [22].

$$CpRe(CO)_2(NO)+BF4-+KOH\ (aq) \rightarrow CpRe(CO)(NO)(COOH)+KBF4$$

$$CpRe(CO)(NO)(COOH)+Ph_3SnCl+Na_2CO_3 \rightarrow$$

$$CpRe(CO)(NO)(CO_2)SnPh_3+NaCl+CO_2+H_2O$$

$$CpRe(CO)(NO)(COOH)+Re(CO)_4(PPh_3)(FBF_3)+Na_2CO_3 \rightarrow$$

$$CpRe(CO)(NO)(CO_2)Re(CO)_3(PPh_3)+NaBF_4+CO_2+H_2O$$

Scheme 2. Synthetic route of rhenium complexes.

Applications of Rhenium-Based Radiopharmaceuticals

The flexibility of rhenium coordination compounds has permitted their research as radiopharmaceuticals for both diagnostic and therapeutic purposes. Rhenium-based imaging agents, including those tagged with the radioactive isotope ^{188}Re, have demonstrated their promise in many nuclear imaging modalities, including single-photon emission computed tomography and positron emission tomography

[17]. Furthermore, the radioactive decay of ^{188}Re and ^{186}Re has been used for targeted cancer therapy, providing intriguing alternatives to established treatment methods. Rhenium-based radiopharmaceuticals can be tailored to specific features, including biodistribution, pharmacokinetics, and targeting, with careful ligand design and functionalization. Rhenium coordination compounds have demonstrated promising outcomes in both preclinical and clinical studies, establishing themselves as a rising category of radiopharmaceuticals with various applications in nuclear medicine. The isotopes of rhenium possess favorable nuclear characteristics, such as suitable half-lives and emission properties, which make them particularly effective for imaging and therapeutic uses. In the realm of diagnostic imaging, rhenium-based compounds are being studied as potential radiotracers for positron emission tomography and single-photon emission computed tomography. The medicinal uses of rhenium coordination complexes have also been extensively explored. Some rhenium compounds are under investigation as potential anticancer agents due to their capacity to induce cytotoxic effects and trigger apoptosis in cancerous cells. Furthermore, the incorporation of rhenium compounds into cellular imaging probes shows potential for diagnostic purposes. Beyond diagnostics, rhenium coordination compounds are also being explored for their therapeutic possibilities [23]. Rhenium coordination compounds have been investigated for a wide range of radiopharmaceutical uses, including the treatment of bone metastases, neuroendocrine tumors, and other cancers. One of the primary benefits of employing rhenium-based radiopharmaceuticals is the possibility to combine the radionuclide into specific delivery systems like peptides, antibodies, or small molecules. For example, rhenium-^{186} and rhenium-^{188} have been conjugated to a variety of bisphosphonate compounds, including rhenium-^{186} etidronate and rhenium-^{188} HEDP, to treat bone metastases. These radiopharmaceuticals accumulate selectively in locations with high bone turnover, providing radiation to metastatic disease sites. Another prospective use for rhenium coordination compounds is in the treatment of neuroendocrine tumors. Rhenium^{183}-labeled somatostatin analogs, such as rhenium-^{188} HYNIC-TATE, have shown promising clinical outcomes, highlighting the promise of rhenium-based radiopharmaceuticals for cancer therapy. The adaptability of rhenium coordination chemistry has also allowed for the creation of rhenium-based radiopharmaceuticals for other cancers, such as prostate cancer and some lymphomas [12]. Fig. (**3**) shows the different applications of rhenium-based radiopharmaceuticals.

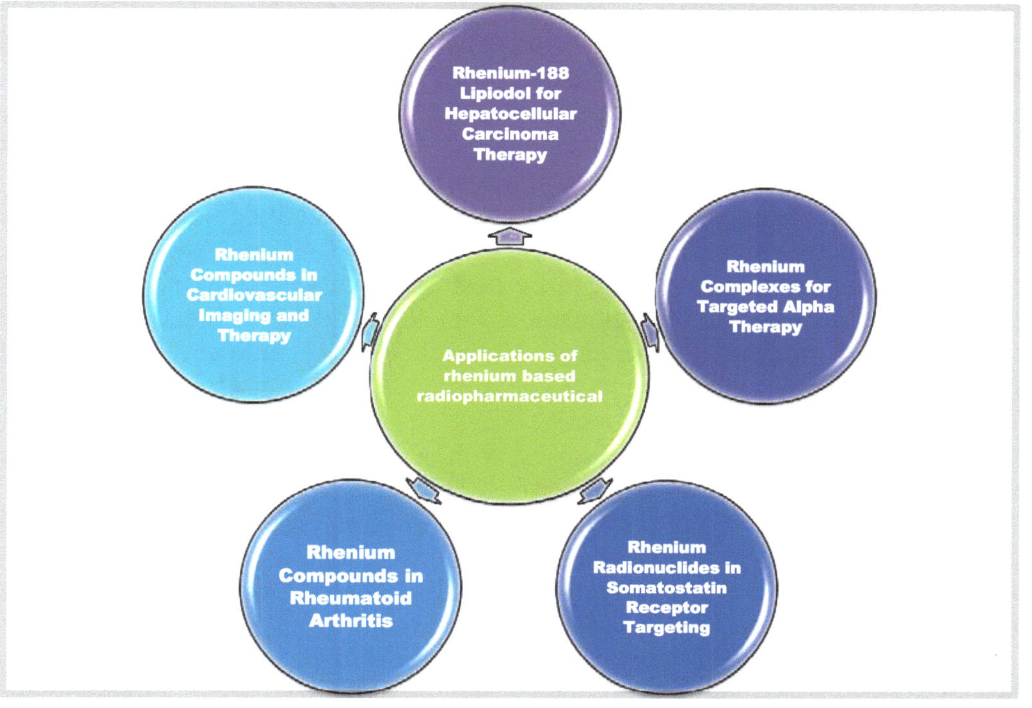

Fig. (3). Different applications of rhenium-based radiopharmaceutical.

Rhenium-Based Radiopharmaceuticals for Bone Pain Palliation

Radiopharmaceuticals are critical in the treatment of metastatic bone pain, especially for patients with advanced cancers such as prostate, breast, and lung cancer. Radiopharmaceuticals are designed to accumulate in bone metastases, producing targeted radiation that relieves pain caused by skeletal lesions. In comparison to conventional medicines, this focused strategy eliminates systemic adverse effects and minimizes damage to adjacent healthy tissues. Most bone-seeking radiopharmaceuticals use beta-emitting radionuclides. They can be divided into two types: inorganic formulations (*e.g.*, ^{89}strontium) that imitate calcium and are absorbed in areas of active bone mineralization, and non-calcium analogue formulations (*e.g.*, ^{153}samarium, ^{186}rhenium) that use phosphonates to bind to calcium in the bone matrix. By efficiently treating pain, these medicines improve the quality of life for individuals suffering from excruciating bone pain caused by metastases. Rhenium isotopes, ^{186}Re and ^{188}Re, produce beta particles. These isotopes emit beta particles after settling in bone metastases. Rhenium emits beta particles with a short range (about a few millimeters), allowing for focused radiation delivery to the metastatic site while sparing surrounding healthy tissues. The radiation from beta particles damages DNA in tumor cells in bone metastases, causing cell death by apoptosis or mitotic catastrophe. This decreases tumor growth and reduces stress on bone structures, thereby reducing pain. This is particularly important for individuals with multiple metastatic lesions, as radiopharmaceuticals tend to be well-tolerated and can provide substantial pain relief [24]. Rhenium compounds, especially ^{186}Re and ^{188}Re, have shown promise in relieving pain caused by bone metastases. These radiopharmaceuticals specifically target bone and have an affinity for areas with heightened osteoblastic activity, which is often observed in metastatic bone diseases. They emit beta radiation, which helps alleviate pain by destroying cancer cells in the bone and lessening the inflammatory responses associated with bone metastases. Initial clinical trials conducted in the early 1990s with ^{186}Re indicated that a single administration of around 33 mCi (1221 MBq) of ^{186}Re hydroxyethylidenediphosphonic acid (HEDP) provided pain relief for roughly 80% of patients with osseous metastases, with only mild and temporary toxicity to the bone marrow noted. In a randomized phase II trial, the effectiveness of 188rhenium was assessed in patients with painful bone metastases. The findings demonstrated that multiple doses of 188Rhenium led to minimal harm to the bone marrow while delivering significant pain relief. Despite their evident effectiveness, rhenium-based radiopharmaceuticals that seek bone have not obtained widespread regulatory approval in the US or Europe, limiting their clinical use [25]. Rhenium radiopharmaceuticals enter the bloodstream and accumulate in bone metastases due to their affinity for bone minerals. They generate beta particles that irradiate and kill tumor cells in bone. Therapeutic radiation reduces tumor activity in the bone, alleviating discomfort (Fig. **4**).

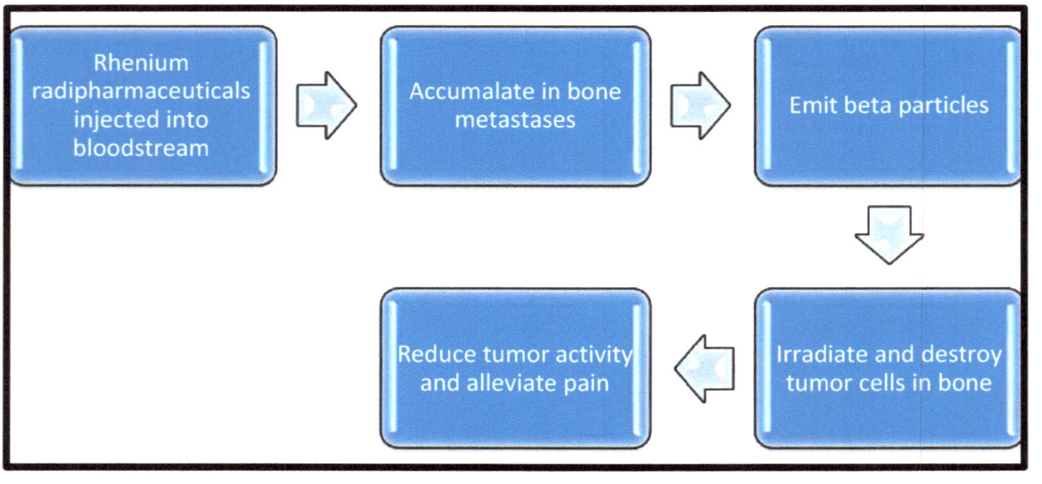

Fig. (4). Mechanism of rhenium-based radiopharmaceuticals for bone pain palliation.

Rhenium Compounds in Targeted Cancer Therapy

Rhenium compounds' distinctive features, such as their structural closeness to platinum-based medications and propensity to cause apoptosis in cancer cells, have prompted much investigation into their potential application as anticancer treatments. Recent research has shown that rhenium compounds can preferentially target and kill cancer cells using a variety of methods, such as DNA damage, oxidative stress, and signaling pathway modification [12]. The discovery of rhenium-based anticancer chemicals represents a viable option for the advancement of targeted cancer therapy, with the potential to overcome the limitations of ^{188}Re. ^{186}Re has also been investigated as a prospective agent for targeted cancer therapy. The ability to bind these rhenium radionuclides to a variety of targeting molecules, including peptides, antibodies, and other biomolecules, enables the selective delivery of therapeutic radiation to cancer cells while sparing healthy tissues. Rhenium compounds have attracted considerable attention in the field of targeted cancer therapy due to their distinct properties and potential mechanisms of action. Various rhenium complexes employ strategies that could induce cytotoxic effects on cancer cells. For example, rhenium(IV) compounds can initiate apoptosis, while rhenium(V) oxo complexes might cause cancer cell death *via* a necroptotic mechanism, which is dependent on the generation of Reactive Oxygen Species (ROS) and the reduction of mitochondrial membrane potential. In comparison to traditional chemotherapeutic agents such as platinum-based drugs, rhenium compounds show greater selectivity for cancer cell lines. Certain rhenium complexes, for instance, have demonstrated significant cytotoxic effects against particular cancer cell lines (like HTB-12 human astrocytoma) while sparing normal cells, indicating a possible decrease in side effects. Moreover, rhenium compounds can be utilized in conjunction with conventional cytotoxic drugs to enhance their overall effectiveness. For instance, combining rhenium complexes with agents such as doxorubicin can capitalize on rhenium's targeting abilities to improve delivery to cancer cells, potentially resulting in better outcomes. The design aspect of rhenium compounds facilitates tailored delivery to cancer cells. For example, the Re(I)-diselenoether complex is noted for its solubility and lipophilicity, which could aid in biodistribution and cellular uptake, making it suitable for targeted therapy. Although numerous rhenium compounds have shown promising results in preclinical experiments, they largely remain within the research phase. This review highlights the necessity for further exploration of their pharmacological and toxicological characteristics to confirm their safety and effectiveness for clinical applications [20]. Rhenium compounds, notably those containing the rhenium(I) tricarbonyl core (Fig. **5**), have been shown to be more potent than cisplatin in certain cancer cell types, including HeLa cells. This suggests that certain rhenium compounds can cause cell death or impede cell proliferation at lower concentrations than cisplatin. The different methods of action of rhenium complexes may contribute to their increased effectiveness, allowing them to target cancer cells that have gained resistance to platinum-based medicines like cisplatin. Rhenium complexes' ability to prevent tumor growth *in vivo* suggests that they could be effective cancer therapies. This is especially essential in the case of ovarian cancer, which frequently poses obstacles such as delayed diagnosis and resistance to traditional treatments. The promising results of both *in vitro* and *in vivo* experiments support the ongoing exploration of rhenium complexes as feasible alternatives or adjuncts to existing cancer treatments, particularly for patients who may not respond well to traditional platinum-based therapy [26] (Table **2**).

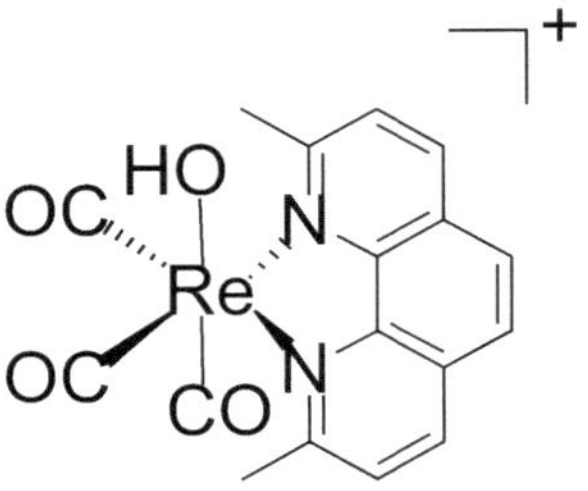

Fig. (5). [RE(CO)$_3$(2,9-dimethyl-1,10-phenanthroline)(OH$_2$)]$^+$.

Table 2. ***In vivo* anticancer properties of the rhenium(I) complex compared to traditional platinum-based medicines.**

Aspects	Rhenium(I) Complexes	Platinum- Based Drugs
Mechanism of Action	Distinct mechanisms that may target mitochondria and cause oxidative stress.	Primarily produces DNA cross-links, which lead to apoptosis.
Potency	In some cancer cell lines, it can be more powerful than cisplatin (for example, HeLa cells).	Although effective, greater doses may be required to get similar benefits.
Resistance	There is no cross-resistance with cisplatin, and it is effective against resistant cancer cells.	Resistance is common, which reduces effectiveness in some patients.
Side effects	Animal models show improved tolerance and potentially fewer side effects.	Significant side effects include nephrotoxicity and nausea.
Biodistribution	Good biodistribution with renal and hepatobiliary excretion; stable *in vivo*.	Variable biodistribution can result in buildup in non-target tissues.
Tumorgrowth inhibition	*In vivo* inhibition of tumor growth has been demonstrated, particularly in ovarian cancer models.	Although effective in reducing tumor growth, it may be restricted by resistance and side effects.

(Table 2) cont.....

Aspects	Rhenium(I) Complexes	Platinum- Based Drugs
Clinical application	Emerging as a possible alternative or addition to platinum-based therapy, further research is needed.	Despite being established as a first-line treatment for a variety of malignancies, resistance remains an issue.
IC_{50} values	HeLa cells: ~5 µM, A2780 ovarian carcinoma: ~7µM	HeLa cells: ~12-15µM, A2780: ~18-20µM
Survival Benefits	Median survival ~2x control, significant extension compared to cisplatin	Survival improved *vs* control, but was limited by resistance
Toxicity	Minimal systemic toxicity, animals maintained their weight and normal behaviour	High nephrotoxicity and systemic toxicity at therapeutic doses

Rhenium-labeled radiopharmaceuticals bind to cancer-specific receptors. Cancer cells ingest the chemical by endocytosis. ^{186}Re and ^{188}Re emit beta particles, which harm cancer cells. Beta particles disrupt DNA strands and produce Reactive Oxygen Species (ROS). The gamma camera analyzes biodistribution (^{186}Re only). Continuous beta radiation causes tumor shrinking. Rhenium is eliminated by the kidneys and liver (Fig. **6**).

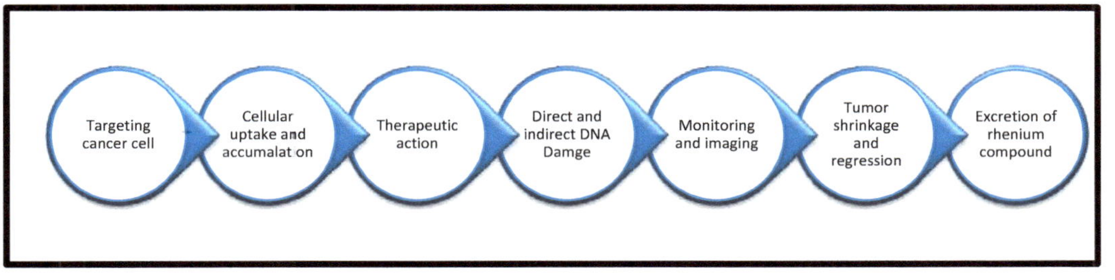

Fig. (6). Mechanism of targeted cancer therapy.

Rhenium Compounds in Hepatocellular Carcinoma

Rhenium compounds, particularly ^{186}Re-idarubicin, have demonstrated promising results in the treatment of Hepatocellular Carcinoma (HCC). The most prevalent type of primary liver cancer is Hepatocellular Carcinoma (HCC), which originates in the liver's main functioning cells, hepatocytes. It is a major global health concern, with high fatality rates. HCC is the third biggest cause of cancer-related deaths globally. It is more common in areas with a high prevalence of chronic liver illness, especially in East Asia and Sub-Saharan Africa. ^{186}Re, a medium-energy β-emitter, improves idarubicin's cytotoxicity through radiotoxicity rather than classical chemotherapeutic action. This is especially useful in targeting cancer cells while limiting damage to nearby healthy organs. The biodistribution of ^{186}Re-idarubicin was assessed in both normal and tumor-induced mice. The findings showed a quick distribution phase with high buildup in the liver, which is important because idarubicin is largely eliminated through this organ. The tracer had substantial initial activity in the blood and liver, indicating successful targeting of HCC. The good biodistribution profile, combined with a high target-to-non-target ratio, implies that ^{186} Re-idarubicin could be used as both a diagnostic and therapeutic agent for HCC. The ability to visualize the compound's dispersion using gamma cameras increases its utility in clinical settings. ^{186}Re-idarubicin has greater cytotoxic effects than cold idarubicin. This is mostly owing to the radiotoxicity of the ^{186}Re isotope, which adds to its greater potential to induce cell death than the non-radioactive form (Table **3**).

Table 3. Comparison btw ^{186}Re-idarubicin and cold idarubicin.

Aspects	^{186}Re- idarubicin	Cold idarubicin
Enhanced cytotoxicity	^{186}Re's radiotoxicity causes increased cytotoxic consequences.	Cytotoxicity is exclusively based on chemotherapeutic activity.
Mechanism of action	Combines chemotherapeutic and radiation actions.	Primarily functions *via* intercalating DNA and inhibiting topoisomerase II.
Dose-Dependent Effects	Inhibits cancer cell proliferation in a dose-dependent manner, with stronger effects at higher dosages.	Exhibits dose-dependent effects, but no additional radiotoxicity.
IC_{50}values	Lower IC_{50} values imply greater potency; less concentration is required for identical effects.	Higher IC_{50} value than ^{186}Re-idarubicin, requiring a higher dose to have identical effects.
Clinical implications	Potentially more effective for treating Hepatocellular Carcinoma (HCC), with better results.	While effective, there may be limits compared to radiolabeled solutions.

Idarubicin, a chemotherapeutic drug, is radiolabeled with ^{186}Re to create the ^{186}Re-idarubicin complex. This requires rhenium^{186} attaching to the idarubicin molecule under ideal conditions (for example, specific pH and concentrations). The ^{186}Re-idarubicin compound is normally given to the patient *via* injection. It circulates throughout the body and is found in malignant tissues, particularly in the liver for Hepatocellular Carcinoma (HCC). Cancer cells absorb the ^{186}Re-idarubicin combination using a variety of methods, including passive diffusion and active transport. The metabolic activity of the cancer cell, as well as the existence of particular transporters, influences absorption. Idarubicin enters cancer cells and intercalates into DNA strands. This means that the idarubicin molecule inserts itself between the bases of the DNA helix, altering the DNA's normal structure and function. Idarubicin inhibits topoisomerase II, an enzyme required for DNA replication and repair. By stabilizing the DNA-enzyme complex, idarubicin prevents the enzyme from resealing the cleaved DNA strands, resulting in the buildup of DNA breaks. The ^{186}Re isotope emits beta radiation, which contributes to its cytotoxic properties. As the radioactive decay occurs, the released radiation causes more damage to the DNA and cellular structures, hence increasing the complex's cytotoxicity. The combined effects of DNA damage from intercalation and radiotoxicity cause cellular stress, which eventually leads to programmed cell death (apoptosis). Cancer cells are unable to repair significant DNA damage, which causes cell death. The targeted delivery of ^{186}Re-idarubicin to cancer cells, together with its dual mechanism of action (chemotherapeutic and radiotoxic), results in efficient tumor growth suppression and probable cancer regression, notably in cases of HCC.

The cytotoxic mechanism of ^{186}Re-idarubicin begins with the creation of a compound containing idarubicin and the radioactive isotope ^{186}Re. Once injected, this compound travels through the circulation and preferentially targets malignant organs, notably the liver. Cancer cells absorb the ^{186}Re-idarubicin combination *via* diffusion and active transport. Idarubicin intercalates into DNA strands within cells, causing structural and functional disruptions. It also inhibits the enzyme topoisomerase II, preventing DNA breaks from being repaired and causing DNA damage to accumulate. The beta radiation released by ^{186}Re contributes to cellular damage, hence increasing the overall cytotoxic effects. The combination of DNA damage and radiotoxicity induces programmed cell death (apoptosis) in cancer cells. Ultimately, ^{186}Re-idarubicin's targeted delivery and dual action result in effective tumor growth suppression and probable cancer regression. Rhenium complexes have anticancer actions *via* numerous interrelated metabolic mechanisms. Rhenium conjugates, such as ^{186}Re-idarubicin, disturb mitochondrial membrane potential, causing cytochrome c release and caspase-dependent cascade activation. In addition, rhenium(V) oxo species have been shown to accelerate the production of Reactive Oxygen Species (ROS), resulting in an oxidative imbalance that destroys cellular lipids, proteins, and nucleic acids. Rhenium complexes can also interact with DNA by intercalating it or by stopping enzymes like topoisomerase II from working, which stabilizes DNA strand breaks and gets in the way of replication and repair. The DNA Damage Response (DDR) is triggered by this accumulating damage, which stops the cell cycle at the G2/M checkpoint. Moreover, the radiotoxic contribution of the ^{186}Re isotope augments these cytotoxic mechanisms, as the released β-particles induce localized ionization events that exacerbate DNA double-strand breaks and oxidative damage. Mitochondrial failure, ROS-mediated stress, and DNA damage all work together to cause cell death. This is why rhenium-based drugs like ^{186}Re-idarubicin are so good at fighting cancer (Table **4**) (Fig. 7) [27 - 29].

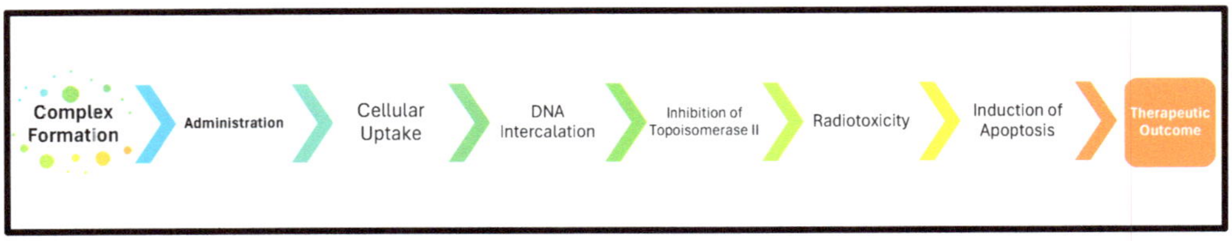

Fig. (7). Mechanism of rhenium compounds in hepatocellular carcinoma.

Table 4. Clinical and preclinical investigations of Rhenium compounds.

Compound isotope	Modal/Target	Phase	Outcomes
^{186}Re-BMEDA liposome (Obisbemeda)	Recurrent glioblastoma (CED delivery)	Phase I dose-escalation (21 pts)	MTD not reached, mOS ~ 11 mo (*vs* ~8 mo SOC), mostly stable disease, safe [30]
^{188}Re-resin (Rhenium SCT)	Basal/Squamous cell carcinoma(NMSC)	Prospective Phase IV (EPIC-Skin, 182 pts)	97.2% lesions CR at 6 mo, no pain, 15.9% mild Aes, safe and effective [31]
^{188}Re-resin brachytherapy	Refractory/difficult BCC(skin)	Retrospective (64 pts)	~93% lesion control at 2 yr, noninvasive, well-tolerated, efficacious [32]
PSMA-GCK01(theranostic ligand)	Metastatic prostate cancer (PSMA)	Preclinical + compassionate use (3pts)	High tumor uptake in mice, human SPECT/Therapy matched 177Lu-PSMA pattern, stable labeling [33]
^{188}Re-Liposome (BMEDTA)	Various tumors (colon, *etc*)	preclinical	High tumor accumulation, strong tumor suppression, and survival benefits, promising radionanotherapy [34]

CHALLENGES

Despite their positive preclinical and early clinical outcomes, rhenium-based radiopharmaceuticals encounter significant obstacles on their way to broad clinical use. Scalability of manufacturing is a significant challenge, as both ^{186}Re and ^{188}Re require stable generator systems or reactor-based synthesis, which might not be available in all clinical sites. This raises questions regarding the supply chain's reliability and isotope availability for multicenter trials. The large-scale production of rhenium-labeled agents, including sterile handling, quality control, and GMP compliance, can be costly compared to more conventional radionuclides like ^{99m}Tc or ^{177}Lu. From a regulatory standpoint, approval routes are complex and vary by area. Rhenium radiopharmaceuticals necessitate stringent toxicological, dosimetric, and long-term safety evaluations; yet, only a limited number of late-phase clinical trials have been conducted thus far. Rhenium radiopharmaceuticals necessitate stringent toxicological, dosimetric, and long-term safety evaluations; yet, only a limited number of late-phase clinical trials have been conducted thus far. Furthermore, the absence of universally accepted guidelines for novel theranostic agents can hinder the approval process. Additionally, challenges in clinical translation involve maintaining consistent radio labeling yields, improving pharmacokinetics for human applications, and demonstrating that the new treatment is at least as effective as, or better than, existing standard care options. To facilitate the extensive application of rhenium-based radiopharmaceuticals in oncology and nuclear medicine, collaboration among university researchers, industry stakeholders, and regulatory bodies is essential to resolve these issues.

CONCLUSION

Rhenium complexes: antitumor potential and applications in bioinorganic chemistry. It underlines that rhenium compounds, particularly those in the +1 and +5 oxidation states, have significant cytotoxicity against a variety of cancer cell lines, making them interesting candidates for future research in cancer therapy. Rhenium-based radiopharmaceuticals are a significant improvement in nuclear medicine, combining the unique features of rhenium complexes with novel targeting tactics and imaging capabilities. Their dual functionality in the ability to serve both diagnostic and therapeutic purposesmakes them essential instruments in the fight against cancer and other disorders. The review emphasizes the role of structural characteristics like ligand type and coordination geometry in determining the biological activity and stability of these complexes. Rhenium's ability to form stable complexes with a variety of ligands lends itself to the development of radiopharmaceuticals, which allow for targetedadministration to malignant regions while limiting damage to healthy cells. The synthesis and characterisation of novel rhenium(I) complexes show that they have the potential to be excellent fluorescence imaging agents, with features such as increased fluorescence when bound to biological cells and compatibility with biological systems. The therapeutic potential of some rhenium isotopes expands their utility, allowing for the investigation of dual diagnostic and therapeutic applications.

FUTURE PERSPECTIVE

Rhenium's distinct oxidation states and coordination chemistry make it an ideal option for creating new catalysts for a variety of chemical reactions, including oxidation, hydrogenation, and C-C coupling reactions. Future research could focus on developing rhenium-based catalysts that are more efficient, selective, and ecologically benign. Rhenium catalysts could be enhanced for use in green chemical applications, lowering waste and energy consumption in industrial operations. Creating new drug delivery vehicles, such as nanoparticles or liposomes, could improve the bioavailability and targeted administration of rhenium complexes to tumour locations. This strategy may reduce systemic toxicity while increasing therapeutic efficacy. Beyond anticancer activities, rhenium compounds may have advantages in imaging and diagnostics due to their distinctive spectroscopic features. This dual capability could provide valuable resources for both cancer treatment and monitoring its progression. The potential use of rhenium compounds alongside current chemotherapeutics or targeted therapies will be an important focus of research. Investigations may focus on synergistic interactions that enhance the overall effectiveness of treatment regimens, particularly for cancers that are resistant to conventional therapies. There is growing interest in combining radionuclide therapy with other treatment modalities, including chemotherapy, immunotherapy, and bisphosphonates. These combinations could improve overall treatment efficacy, patient outcomes, and help overcome resistance mechanisms in cancer cells. The future of radionuclide therapy might involve more personalized approaches, with treatment strategies customized to specific patient characteristics such as tumor biology, genetic markers, and unique metastasis patterns. This could result in more effective and targeted treatment methods. The advancement of innovative imaging techniques, such as hybrid imaging systems (*e.g.*, PET/MRI), will enhance the applicability of rhenium-based radiopharmaceuticals. Enhanced imaging capabilities will allow for better visualization of tumor behavior, responses to treatment, and the pharmacokinetics of radiopharmaceuticals, leading to more informed clinical choices.

REFERENCES

[1] Romão CC, Kühn FE, Herrmann WA. Rhenium (VII) oxo and imido complexes: synthesis, structures, and applications. Chem Rev 1997; 97(8): 3197-246. [http://dx.doi.org/10.1021/cr9703212] [PMID: 11851489]

[2] Paes FM, Serafini AN. Systemic metabolic radiopharmaceutical therapy in the treatment of metastatic bone pain. Semin Nucl Med 2010; 40(2): 89-104. [http://dx.doi.org/10.1053/j.semnuclmed.2009.10.003] [PMID: 20113678]

[3] Dash A, Das T. Production of therapeutic radionuclides for bone-seeking radiopharmaceuticals to meet clinical exigency and academic research. J Radioanal Nucl Chem 2025; 334(4): 2591-611. [http://dx.doi.org/10.1007/s10967-025-10025-1]

[4] Kostova I. Lanthanides as anticancer agents. Curr Med Chem Anticancer Agents 2005; 5(6): 591-602. [http://dx.doi.org/10.2174/156801105774574694] [PMID: 16305481]

[5] Melis DR, Burgoyne AR, Ooms M, Gasser G. Bifunctional chelators for radiorhenium: past, present and future outlook. RSC Med Chem 2022; 13(3): 217-45. [http://dx.doi.org/10.1039/D1MD00364J] [PMID: 35434629]

[6] Kumar A, Bal C, Srivastava DN, *et al.* Management of multiple intrahepatic recurrences after radiofrequency ablation of hepatocellular carcinoma with rhenium-188-HDD-lipiodol. Eur J Gastroenterol Hepatol 2006; 18(2): 219-23. [http://dx.doi.org/10.1097/00042737-200602000-00016] [PMID: 16394805]

[7] Hashimoto K. Synthesis of a 188Re-HEDP complex using carrier-free 188Re, and a study of its stability. Appl Radiat Isot 1998; 49(4): 351-6. [http://dx.doi.org/10.1016/S0969-8043(97)00284-4]

[8] Dilworth JR. Rhenium chemistry–then and now. Coord Chem Rev 2021; 436: 213822. [http://dx.doi.org/10.1016/j.ccr.2021.213822]

[9] Lepareur N, Lacœuille F, Bouvry C, *et al.* Rhenium-188 labeled radiopharmaceuticals: current clinical applications in oncology and promising perspectives. Front Med (Lausanne) 2019; 6: 132. [http://dx.doi.org/10.3389/fmed.2019.00132] [PMID: 31259173]

[10] Makris G, Radford LL, Kuchuk M, *et al.* NOTA and NODAGA [99mTc] Tc-and [186Re] Re-tricarbonyl complexes: radiochemistry and first example of a [99mTc] Tc-NODAGA somatostatin receptor-targeting bioconjugate. Bioconjug Chem 2018; 29(12): 4040-9. [http://dx.doi.org/10.1021/acs.bioconjchem.8b00670] [PMID: 30412382]

[11] Boschi A, Martini P, Uccelli L. 188Re (V) nitrido radiopharmaceuticals for radionuclide therapy. Pharmaceuticals (Basel) 2017; 10(1): 12. [http://dx.doi.org/10.3390/ph10010012] [PMID: 28106830]

[12] Martínez-Lillo J, Mastropietro TF, Lappano R, *et al.* Rhenium(iv) compounds inducing apoptosis in cancer cells. Chem Commun (Camb) 2011; 47(18): 5283-5. [http://dx.doi.org/10.1039/c1cc11038a] [PMID: 21448489]

[13] Deutsch E, Libson K, Vanderheyden JL, Ketring AR, Maxon HR. The chemistry of rhenium and technetium as related to the use of isotopes of these elements in therapeutic and diagnostic nuclear medicine. Int J Rad Appl Instrum B 1986; 13(4): 465-77. [http://dx.doi.org/10.1016/0883-2897(86)90027-9] [PMID: 3793504]

[14] Lengacher R, Marlin A, Śmiłowicz D, Boros E. Medicinal inorganic chemistry – challenges, opportunities and guidelines to develop the next generation of radioactive, photoactivated and active site inhibiting metal-based medicines. Chem Soc Rev 2022; 51(18): 7715-31. [http://dx.doi.org/10.1039/D2CS00407K] [PMID: 35942718]

[15] White JM, Escorcia FE, Viola NT. Perspectives on metals-based radioimmunotherapy (RIT): moving forward. Theranostics 2021; 11(13): 6293-314. [http://dx.doi.org/10.7150/thno.57177] [PMID: 33995659]

[16] Hu A, Wilson JJ. The Radiopharmaceutical Chemistry of Metallic Radionuclides In Radiopharmaceutical Therapy. Cham: Springer International Publishing, 2023; 19: (pp. 123-144). [http://dx.doi.org/10.1007/978-3-031-39005-0_6]

[17] Huang Z, Wilson JJ. Therapeutic and diagnostic applications of multimetallic rhenium (I) tricarbonyl complexes. Eur J Inorg Chem 2021; 2021(14): 1312-24. [http://dx.doi.org/10.1002/ejic.202100031]

[18] Gothe ML, Silva KLC, Figueredo AL, *et al.* Rhenium–a tuneable player in tailored hydrogenation catalysis. Eur J Inorg Chem 2021; 2021(39): 4043-65. [http://dx.doi.org/10.1002/ejic.202100459]

[19] Archer CM, Dilworth JR, Griffiths DV, *et al.* Technetium and rhenium oxo-complexes of new tetradentate ligands with N 2 S 2 and NS 3 donor sets. J Chem Soc, Dalton Trans 1997; (8): 1403-10. [http://dx.doi.org/10.1039/a606112e]

[20] Collery P, Desmaele D, Vijaykumar V. Design of rhenium compounds in targeted anticancer therapeutics. Curr Pharm Des 2019; 25(31): 3306-22. [http://dx.doi.org/10.2174/1381612825666190902161400] [PMID: 31475892]

[21] Gerber TIA, Hosten E, Mayer P, Tshentu ZR. Synthesis and characterization of rhenium(III) and (V) pyridylimidazole complexes. J Coord Chem 2006; 59(3): 243-53. [http://dx.doi.org/10.1080/00958970500069063]

[22] Gibson DH, Mehta JM, Ye M, Richardson JF, Mashuta MS. Synthesis and characterization of rhenium metallocarboxylates. Organometallics 1994; 13(4): 1070-2. [http://dx.doi.org/10.1021/om00016a008]

[23] Lo KKW, Zhang KY, Li SPY. Recent exploitation of luminescent rhenium (I) tricarbonyl polypyridine complexes as biomolecular and cellular probes. Eur J Inorg Chem 2011; 2011(24): 3551-68. [http://dx.doi.org/10.1002/ejic.201100469]

[24] Das T, Banerjee S. Radiopharmaceuticals for metastatic bone pain palliation: available options in the clinical domain and their comparisons. Clin Exp Metastasis 2017; 34(1): 1-10. [http://dx.doi.org/10.1007/s10585-016-9831-9] [PMID: 27988895]

[25] Sartor O, Hoskin P, Bruland ØS. Targeted radio-nuclide therapy of skeletal metastases. Cancer Treat Rev 2013; 39(1): 18-26. [http://dx.doi.org/10.1016/j.ctrv.2012.03.006] [PMID: 22534284]

[26] Konkankit CC, King AP, Knopf KM, Southard TL, Wilson JJ. *In vivo* anticancer activity of a rhenium (I) tricarbonyl complex. ACS Med Chem Lett 2019; 10(5): 822-7. [http://dx.doi.org/10.1021/acsmedchemlett.9b00128] [PMID: 31098006]

[27] El-Kawy OA, Talaat HM. Preparation, characterization and evaluation of ^{186} Re-idarubicin: a novel agent for diagnosis and treatment of hepatocellular carcinoma. J Labelled Comp Radiopharm 2016; 59(2): 72-7. [http://dx.doi.org/10.1002/jlcr.3368] [PMID: 26725469]

[28] Huang FYJ, Lee TW, Kao CHK, *et al.* Imaging, autoradiography, and biodistribution of (188)Re-labeled PEGylated nanoliposome in orthotopic glioma bearing rat model. Cancer Biother Radiopharm 2011; 26(6): 717-25. [http://dx.doi.org/10.1089/cbr.2011.1052] [PMID: 22145660]

[29] Kaushik D, Jangra P, Verma R, *et al.* Radiopharmaceuticals: An insight into the latest advances in medical uses and regulatory perspectives. J Biosci 2021; 46(1): 27. [http://dx.doi.org/10.1007/s12038-021-00147-5] [PMID: 33753578]

[30] Brenner AJ, Patel T, Bao A, *et al.* Convection enhanced delivery of Rhenium (^{186}Re) Obisbemeda (^{186}RNL) in recurrent glioma: a multicenter, single arm, phase 1 clinical trial. Nat Commun 2025; 16(1): 2079. [http://dx.doi.org/10.1038/s41467-025-57263-1] [PMID: 40055350]

[31] Baxi S, Vohra S, Hong A, *et al.* Effectiveness and patient experiences of rhenium skin cancer therapy for nonmelanoma skin cancer: interim results from the EPIC-Skin study. J Nucl Med 2024; 65(9): 1450-5. [http://dx.doi.org/10.2967/jnumed.124.267988] [PMID: 39025650]

[32] Chessa MA, Baraldi C, Savoia F, *et al.* Unsealed ^{188}Rhenium resin brachytherapy in non-surgical candidates with refractory basal cell carcinoma: clinical outcomes. Dermatol Pract Concept 2025; 15(2): 4993. [http://dx.doi.org/10.5826/dpc.1502a4993] [PMID: 40401867]

[33] Cardinale J, Giesel FL, Wensky C, Rathke HG, Haberkorn U, Kratochwil C. PSMA-GCK01: a generator-based 99mTc/188Re theranostic ligand for the prostate-specific membrane antigen. J Nucl Med 2023; 64(7): 1069-75. [http://dx.doi.org/10.2967/jnumed.122.264944] [PMID: 36759199]

[34] Chang CH, Chang MC, Chang YJ, Chen LC, Lee TW, Ting G. Translating research for the radiotheranostics of nanotargeted 188Re-liposome. Int J Mol Sci 2021; 22(8): 3868. [http://dx.doi.org/10.3390/ijms22083868] [PMID: 33918011]

SUBJECT INDEX

A

B

C

Abhay Nanda Srivastva (Ed.)
All rights reserved-© 2026 Bentham Science Publishers

D

E

F

G

H

I

K

L

M

N

O

P

Q

R

S

T

U

V

W

X

Z

www.ingramcontent.com/pod-product-compliance
Lightning Source LLC
LaVergne TN
LVHW070951210926
848482LV00002B/172